A

Principles of Linear Algebra

Principles of Linear Algebra

J. LARRIEU

Institut de Statistique
Université de Paris

Translated from the French by
Marina SMITH-KOK

GORDON AND BREACH
Science Publishers
New York London Paris

First published in French as *Principes d'algèbre linéaire* by Dunod

Editorial office for the United Kingdom:

Gordon and Breach, Science Publishers Ltd.
41/42 William IV Street
London WC2

Editorial office for France:

Gordon & Breach
7-9 rue Emile Dubois
Paris 14e

Library of Congress catalog card number 72-109835. ISBN 0 677 01610 7 (cloth); 0 677 01615 8 (paper). Printed in Great Britain.

CONTENTS

PREFACE

This work contains the basic elements of linear algebra from which most properties of matrix calculus can be derived very naturally. Classical methods are used whenever possible so that a good knowledge of elementary calculus is all that is necessary to follow the reasoning. The first chapter contains however a general exposition of vector spaces in modern terminology, so as to promote a clear understanding of concepts which might otherwise create real difficulties at a later stage.

Chapter VIII is at a somewhat higher level than the rest of the book as we wanted to give a thorough account of methods which are so useful in so many situations.

We urge the reader to make what little effort is necessary to master the use of matrices; it is well worth the trouble and Chapter X is entirely devoted to showing the importance of linear algebra in the representation and the analysis of models which would be almost inextricable without it.

CHAPTER I

GENERALITIES ON VECTOR SPACES

In this chapter we adopt a rather general point of view so as to give the widest possible range of application to the results we will derive. This will not complicate our analysis, on the contrary.

We will be dealing with sets E of elements V defined over a field K; we first recall a few definitions.

A field K consists of a finite or infinite set of elements A with two operations defined over these elements.

Each of these operations associates an element of K to every pair (a,b), $a \in K$ ([1]), $b \in K$. These operations will be denoted + and $\times$.

The sets E will be constructed from elements of K in a manner we shall indicate.

In other words, we are going to consider sets E of elements V over which we define operations and relations compatible with the fields K used to define them. We therefore introduce operations involving the elements of E and those of K simultaneously.

These operations have the following fundamental properties which are set as axioms and will constantly be used in the sequel:

([1]) The symbol $\in$ means "belongs to."

OPERATIONS DEFINED OVER THE FIELDS K AND THEIR PROPERTIES

We define over the field K two operations represented by the symbols + and × which associate to every pair of elements (a,b), $a,b \in K$, an element c.

PROPERTIES OF THE + OPERATION

The + operation is closed, i.e.:

(I-1) $$a + b \in K$$

for any $a, b \in K$.

It is associative:

(I-2) $$(a + b) + c = a + (b + c)$$

for any $a, b, c \in K$.

It is commutative:

(I-3) $$a + b = b + a$$

for any $a, b \in K$.

There exists a neutral element called zero (0) such that:

(I-4) $$a + 0 = 0 + a = a$$

for any $a \in K$.

Every element of K has an inverse with respect to the + operation, that is, to every element $a \in K$ there corresponds an element $(-a) \in K$ such that:

(I-5) $$a + (-a) = 0\,.$$

PROPERTIES OF THE $\times$ OPERATION

The $\times$ operation is closed, i.e.:

(I-6) $$a \times b \in K$$

for any $a, b \in K$.

It is associative:

(I-7) $$(a \times b) \times c = a \times (b \times c)$$

for any $a, b, c \in K$.

It is commutative:

(I-8) $$a \times b = b \times a$$

for any $a, b \in K$.

It is distributive with respect to the + operation, i.e.:

(I-9) $$a \times (b + c) = (a \times b) + (a \times c)$$

for any $a, b, c \in K$.

There exists a neutral element called one (1) such that

(I-10) $$a \times 1 = 1 \times a = a$$

for any $a \in K$,

$$1 \in K.$$

Every element of K except the neutral element with respect to the + operation has an inverse with respect to the $\times$ operation, that is, to every element $a \in K$ there corresponds an element $1/a \in K$ such that:

(I-11) $$a \times \frac{1}{a} = 1, \text{ except for } 0.$$

N.B. The properties of fields are clearly generalizations of the usual algebraic properties of numbers.

OPERATION DEFINED OVER THE SET *E* AND IT PROPERTIES

We define over the set *E* an operation $\oplus$ which has exactly the same properties as the + operation defined over the field *K*. We recall these properties briefly:

The $\oplus$ operation is closed:

(I-12) $$V_1 \oplus V_2 \in E, \text{ if } V_1, V_2 \in E.$$

It is associative

(I-13) $$(V_1 \oplus V_2) \oplus V_3 = V_1 \oplus (V_2 \oplus V_3) \qquad V_1, V_2, V_3 \in E.$$

It is commutative:

(I-14) $$V_1 \oplus V_2 = V_2 \oplus V_1 .$$

There exists a neutral element called zero (θ) such that

(I-15) $$V \oplus \theta = V \qquad V \in E, \quad \theta \in E.$$

Every element of E has an inverse with respect to θ, that is, to every element $V \in E$ there corresponds an element $(-V) \in E$ such that

(I-16) $$V \oplus (-V) = \theta .$$

OPERATION DEFINED OVER THE FIELD *K* AND OVER THE SET *E*

We define over the field *K* and the set *E* an operation $\otimes$ which associates a new element to every pair (a, V).

PROPERTIES OF THE $\otimes$ OPERATION

The $\otimes$ operation is closed with respect to the set E, that is:

(I-17) $$a \otimes V \in E$$

for any $a \in K$, $V \in E$.

It is distributive with respect to the $\oplus$ operation defined in E, that is:

(I-18) $$a \otimes (V_1 \oplus V_2) = a \otimes V_1 \oplus a \otimes V_2, \quad a \in K, \quad V_1, V_2 \in E.$$

It is distributive with respect to the + operation defined in K, that is:

(I-19) $$(a + b) \otimes V = a \otimes V \oplus b \otimes V, \quad a, b \in K, \quad V \in E.$$

It is associative with respect to the $\times$ operation defined in K, that is:

(I-20) $$a \otimes (b \otimes V) = (a \times b) \otimes V.$$

The neutral element with respect to the $\times$ operation in the field K is also the neutral element with respect to the $\otimes$ operation, that is:

(I-21) $$1 \otimes V = V$$

for any $V \in E$.

The set of elements $V \in E$ and $a \in K$ associated with the operations, +, $\times$, $\oplus$, $\otimes$ defined above form a vector space over the field K.

These definitions may seem abstract; in reality, they cover very familiar concepts. A very simple example of a vector space is the set of three-dimensional vectors defined in classical analytic geometry over the real numbers. We have merely formalized and axiomatized concepts which seem quite natural in their concrete use. This effort at abstraction is not without a price, for we will later meet results which might offend "common sense" but are rich in possibilities.

The definitions given above allow us to prove a few properties which will constantly be used in the sequel.

Theorems. The following relations are valid:

(I-22) $$0 \otimes V = \theta$$

for any $V \in E$,

(I-23) $$(-1) \otimes V = -V$$

for any $V \in E$,

(I-24) $$a \otimes \theta = \theta$$

for any $a \in K$.

To prove the first, we consider the following relation

$$(0 + 1) \otimes V = 0 \otimes V \oplus 1 \otimes V = 1 \otimes V.$$

We have thus

$$0 \otimes V = \theta.$$

To prove the second, we write:

$$(-1 + 1) \otimes V = -1 \otimes V \oplus 1 \otimes V = 0 \otimes V = \theta,$$

or

$$-1 \otimes V + 1 \otimes V = -1 \otimes V \oplus V = \theta.$$

The third relation can be expressed as

$$a \otimes (\theta \oplus V) = a \otimes \theta \oplus a \otimes V = a \otimes V,$$

whence:

$$a \otimes \theta = \theta .$$

We have seen above that vector spaces are composed of elements with certain properties. It will often be necessary to consider subsets of a vector space, as we shall see later. We now give some definitions and properties connected with these subsets.

SUBSETS OF A VECTOR SPACE

Consider a vector space E consisting of vectors V and a subset R of this space, that is, a set all elements of which belong to E but such that there exist elements of E not belonging to R.

The subset R is itself a vector space over the field K if it satisfies all the conditions and possesses all the properties stated above.

In reality it is less difficult than it may seem at first glance to verify that a subset R is a vector space. Indeed, since the properties (I-1) to (I-11) are relative to the field K they hold automatically, since the subsets R are considered over the same field K as the initial vector space.

We therefore need only verify that the properties (I-12) to (I-21) are satisfied by the elements of the set R. Now, since these are vectors of the vector space E, they satisfy the relations (I-13), (I-14), (I-18), (I-19), (I-20), and (I-21). We must therefore find out whether the conditions (I-12), (I-15), (I-16), (I-17) are satisfied.

(I-25) *A subset R of E is a vector space if and only if it contains the neutral element θ with* *respect to the $\oplus$ operation and the two operations $\oplus$ and $\otimes$ are closed.*

The fact that the $\oplus$ and the $\otimes$ operations are closed implies that the conditions (I-12) and (I-17) are satisfied. The existence of the element θ in R implies the condition (I-15).

It is then sufficient to show that these three conditions imply the condition (I-16), namely that the vector $-V$ associated with every vector $V \in R$ also belongs to R.

Since the $\otimes$ operation is closed, it follows that $a \otimes V \in R$ if $V \in R$. This is true in particular for $(-1) \otimes V$. Now we have seen earlier (I-23) that $(-1) \otimes V$ is equal to $-V$; hence $-V \in R$.

Furthermore

(I-26) *The intersection $R_1 \cap R_2$ of two vector subspaces of E is a vector subspace.*

We recall that the intersection of two sets A and B is the set of elements which belong to the set A and to the set B; the union of these sets consists of the vectors belonging to A *or* to B (this union is denoted $A \cup B$).

The element θ belongs to $R_1 \cap R_2$ since it belongs to all subspaces -- and hence to R_1 and R_2. Furthermore, if U and V belong to $R_1 \cap R_2$ this means that

$$U \in R_1, \quad V \in R_1,$$
$$U \in R_2, \quad V \in R_2.$$

Hence:

$$U \oplus V \in R_1$$
$$U \oplus V \in R_2,$$

and thus

$$U \oplus V \in R_1 \cap R_2 .$$

The $\oplus$ operation is therefore closed.

If $U \in R_1 \cap R_2$, then $U \in R_1$ and therefore $a \otimes U \in R_1$ Similarly $U \in R_2$ and therefore $a \otimes U \in R_2$, so that

$$a \otimes U \in R_1 \cap R_2 .$$

The $\otimes$ operation is therefore closed also, so that $R_1 \cap R_2$ is indeed a vector space.

On the other hand, *the union of two vector subspaces of E is not in general a vector subspace.*

N.B. From now on and for simplicity, we represent the + and the $\oplus$ operations by the same symbol, and completely omit the symbols $\times$ and $\otimes$. In other words, we write:

$$a_1 \otimes V_1 \oplus a_2 \otimes V_2 = a_1 V_1 + a_2 V_2 .$$

No confusion is possible since we will always denote the elements of the field *K* (scalars) by small letters and the vectors by capital letters, except for exceptions which will be duly indicated.

Consider *k* vectors $V_1, V_2, \ldots, V_k$ in the vector space *E*.

(I-27) *The set of linear combinations of these vectors, that is, the set of elements of the form*

$$A = a_1 V_1 + a_2 V_2 + \cdots + a_k V_k , \quad a_i \in K , \quad V_i \in E ,$$

forms a vector subspace of E; it is the vector subspace $\mathfrak{V}$ *generated by the vectors* V_i.

It follows from the properties (I-12) and (I-17) that A belongs to E, and $\mathcal{V}$ is thus a subset of E. It follows from property (I-22) that θ belongs to $\mathcal{V}$. If moreover $A \in \mathcal{V}$ and $B \in \mathcal{V}$, we have

$$A = \sum_i^k a_i V_i, \qquad B = \sum_i^k b_i V_i$$

and consequently

$$A + B = \sum_1^k (a_i + b_i) V_i \text{ where } a_i + b_i \in K .$$

The + operation is therefore closed.

On the other hand, if $A \in \mathcal{V}$ we have

$$aA = a \sum_1^k a_i V_i = \sum_1^k aa_i V_i, \text{ where } aa_i \in K$$

So that $aA \in \mathcal{V}$ and the $\times$ operation is also closed. It follows that $\mathcal{V}$ is a vector space.

DIMENSION OF A VECTOR SPACE

Consider in a vector space a sequence of vectors $V_1, V_2, \ldots, V_p$ such that it is impossible to have the relation

$$\alpha_1 V_1 + \alpha_2 V_2 + \cdots + \alpha_p V_p = \theta, \quad \alpha_i \in K,$$

unless all the α_i are identically zero.

These vectors are said to be linearly independent. If it is possible to write this relation without all the α_i being zero, *the vectors V_i are linearly dependent.*

This concept of independence is the most fundamental concept of all the theory of linear algebra.

We now prove two theorems concerning linear independence.

(I-28) *If the n vectors* $V_1, V_2, ..., V_n$ *are linearly independent, k vectors* V_i *chosen among these n vectors are always linearly independent.*

If this were not the case we could always find a non-zero scalar among the k scalars α_j such that

$$\sum_1^k \alpha_j V_j = 0,$$

and consequently we could write a linear combination of the vectors V_i whose coefficients are not all zero by completing the above sum by the $n - k$ terms $\alpha_i V_i$ such that $\alpha_i = 0$.

Similarly

(I-29) *If n vectors are linearly dependent, any set of m vectors* $(m > n)$ *containing the n first ones is linearly dependent.*

The proof is analogous to that given above.

Two situations may arise in a vector space E as regards the independence of vectors.

a) There is a finite number n such that it is impossible to find more than n linearly independent vectors.

b) However large n may be, it is always possible to find more than n linearly independent vectors.

In the first case, the vector space under consideration is said to be n-dimensional and is denoted E_n.

In the second case, the space E is said to have infinite dimension and is denoted E_∞.

In this book we consider only vector spaces corresponding to the first case, that is, vector spaces with finite dimension.

BASIS OF A VECTOR SPACE

Consider an n-dimensional vector space E_n and a set of n linearly independent vectors $V_1, V_2, \ldots, V_n$ in E_n. From the definition of the dimension of a vector space, if we add *any* $(n + 1)$th vector $V \in E_n$ to this set of vectors, the set $V_1, V_2, \ldots, V_n, V$ will be linearly dependent, that is, we can find $n + 1$ scalars which are not all zero, say $\alpha_1, \alpha_2, \ldots, \alpha_n, \lambda$ such that

$$\alpha_1 V_1 + \alpha_2 V_2 + \cdots + \alpha_n V_n + \lambda V = \theta,$$

and the coefficient λ is not zero because if it were, one of the other coefficients α_i would be non-zero, which contradicts our hypothesis of independence of the vectors $V_1, V_2, \ldots, V_n$. Using the general properties of vector spaces we can therefore write the above relation in the form:

$$V = \beta_1 V_1 + \beta_2 V_2 + \cdots + \beta_n V_n, \quad \text{where } \beta_i = -\alpha_i/\lambda,$$

and this for any vector $V \in E_n$.

In other words, *any vector V of E_n can be expressed as a linear combination of the vectors V_i.*

The vectors V_i are said to form a basis of the vector space E_n and the coefficients β_i are the components or coordinates of the vector V in this basis.

Conversely, we show that:

(I-30) *If any vector of the space E is a linear combination of the n linearly independent vectors $V_1, V_2, \ldots, V_n$ it is impossible to find more than n linearly independent vectors in the space under consideration.*

Suppose we have found p linearly independent vectors $U_1, U_2, \ldots, U_p$, with $p > n$. By hypothesis, all the vectors of E can be expressed in terms of the vectors V_i and in particular:

$$(A)\begin{cases} U_1 = \alpha_1^1 V_1 + \alpha_2^1 V_2 + \cdots + \alpha_n^1 V_n\,, \\ U_2 = \alpha_1^2 V_1 + \alpha_2^2 V_2 + \cdots + \alpha_n^2 V_n\,, \\ \vdots \\ U_p = \alpha_1^p V_1 + \alpha_2^p V_2 + \ldots + \alpha_n^p V_n\,. \end{cases}$$

We are going to show that it is possible to find coefficients $x_1, x_2, \ldots, x_p$ not all zero such that

$$x_1 U_1 + x_2 U_2 + \cdots + x_p U_p = \theta\,,$$

which will contradict our hypothesis of independence of the U_j.

In the system (A) we multiply both sides of the first equality by x_1, both sides of the second by x_2, etc., and add the members on either side of all the relations; we find

$$\begin{aligned} x_1 U_1 + x_2 U_2 + \cdots + x_p U_p &= \\ = (x_1 \alpha_1^1 + x_2 \alpha_1^2 + \cdots + x_p \alpha_1^p) V_1 &+ (x_1 \alpha_2^1 + x_2 \alpha_2^2 + \cdots + x_p \alpha_2^p) V_2 \\ + \cdots + (x_1 \alpha_n^1 + x_2 \alpha_n^2 + \cdots + x_p \alpha_n^p) V_n\,. \end{aligned}$$

Since the V_i are linearly independent, this expression can be equal to zero only if all the coefficients of the V_i are zero, that is, if we have simultaneously:

$$\begin{aligned} \alpha_1^1 x_1 + \alpha_1^2 x_2 + \alpha_1^3 x_3 + \cdots + \alpha_1^p x_p &= 0\,, \\ \alpha_2^1 x_1 + \alpha_2^2 x_2 + \alpha_2^3 x_3 + \cdots + \alpha_2^p x_p &= 0\,, \\ &\vdots \\ \alpha_n^1 x_1 + \alpha_n^2 x_2 + \alpha_n^3 x_3 + \cdots + \alpha_n^p x_p &= 0\,. \end{aligned}$$

This is a homogeneous ([1]) system of n linear equations in p unknowns $x_1, x_2, \ldots, x_p$ where $p > n$. Solving the system will consist in finding a set of p numbers $x'_1, x'_2, \ldots, x'_p$, such that the given equations are satisfied simultaneously when the x_i are replaced by the x'_i, which are then called solutions of the system.

The above system has the obvious solution

$$x_1 = x_2 = \cdots = x_p = 0.$$

This solution is called *the trivial solution.*

We are going to show that there is a second solution for which all the x_i are not zero. We shall need the following lemma:

(I-31) *Every homogeneous system of linear equations consisting of n equations in p unknowns, where the number of unknowns is greater than the number of equations, admits at least one non-trivial solution.*

We prove this lemma by induction, showing that if it holds for $n - 1$ it must also hold for n. Now it is obviously true for $n = 1$. If there is only one equation, we can take any values for $n - 1$ unknowns x_i. The value of the last one can be deduced from them by means of the equation.

Let us then consider the following system:

([1]) A linear system is said to be homogeneous if its solution still satisfies the system when each unknown x_i is multiplied by a number different from zero, which is the case here.

$$
(B)\begin{cases}
A_1 = \alpha_1^1 x_1 + \alpha_1^2 x_2 + \cdots + \alpha_1^p x_p = 0 \\
A_2 = \alpha_2^1 x_1 + \alpha_2^2 x_2 + \cdots + \alpha_2^p x_p = 0 \\
\vdots \\
A_n = \alpha_n^1 x_1 + \alpha_n^2 x_2 + \cdots + \alpha_n^p x_p = 0
\end{cases}
$$

In the equation $A_1 = 0$, all the coefficients are not identically zero, for if they were the equation would vanish. We may suppose without loss of generality that α_1^1 is not zero and consider the systems

$$
(C)\begin{cases}
A_1 = 0 \\
\left.\begin{aligned}
& A_2 - \frac{\alpha_2^1}{\alpha_1^1} A_1 = 0 \\
& A_3 - \frac{\alpha_3^1}{\alpha_1^1} A_1 = 0 \\
& \vdots \\
& A_n - \frac{\alpha_n^1}{\alpha_1^1} A_1 = 0
\end{aligned}\right\} (D)
\end{cases}
$$

We immediately see that the systems *(B)* and *(C)* are equivalent since every solution of the system *(B)* is a solution of the system *(C)* and every solution of the system *(C)* is a solution of the system *(B)*.

Let us now consider the system *(D)*; it does not contain x_1; it is therefore a system of $n - 1$ equations in $p - 1$ unknowns; we have assumed that the lemma is true for $n - 1$; consequently, there is a set of numbers $X_2, X_3, \ldots, X_p$ not all zero which is a solution of the system *(D)*.

Now, if we take

$$X_1 = -\frac{\alpha_1^2 X_2 + \alpha_1^3 X_3 + \cdots + \alpha_1^p X_p}{\alpha_1^1}$$

we see that the equation $A_1 = 0$ is also satisfied;

hence the system *(B)* does indeed admit a non-trivial solution: $X_1, X_2, \ldots, X_p$.

Direct application of this lemma to the system considered above shows that it is possible to find numbers $x_1, x_2, \ldots, x_p$ not all zero, such that:

$$x_1 U_1 + x_2 U_2 + \cdots + x_p U_p = 0.$$

The U_p are therefore not linearly independent. Whence the theorem:

(I-32) *If all the vectors in a vector space E can be expressed as a linear combination of n linearly independent vectors in this space, the space under consideration is n-dimensional.*

This is a fundamental theorem since it is the theorem which is most used in determining the dimension of a vector space.

We now prove some important theorems which will constantly be used in the sequel.

(I-33) *In an n-dimensional vector space, the representation of a vector V as a linear combination of n linearly independent vectors of E_n is unique.*

We have seen earlier that V could be expressed as follows:

$$V = v_1 V_1 + v_2 V_2 + \cdots + v_n V_n .$$

Suppose it is possible to write V as another linear combination of the V_i:

$$V = u_1 V_1 + u_2 V_2 + \cdots + u_n V_n .$$

Applying the properties of vector spaces, we can write

$$-V = -u_1 V_1 - u_2 V_2 - \cdots - u_n V_n$$

and consequently

$$V - V = \theta = (v_1 - u_1) V_1 + (v_2 - u_2) V_2 + \cdots + (v_n - u_n) V_n.$$

Now, since the vectors V_i are linearly independent, the relation given above is possible only if all the coefficients of the V_i are zero, that is, if

$$u_i = v_i \quad \text{for} \quad i = 1, 2, \ldots, n.$$

The representation is therefore unique.

(I-34) *In an n-dimensional vector space E_n, the set generated by p linearly independent vectors $V_1, V_2, \ldots, V_p$ is a p-dimensional vector subspace of E_n.*

It follows from Theorem (I-27) that it is a vector space, and from Theorem (I-32) that it is p-dimensional.

(I-35) *In an n-dimensional vector space E_n, it is always possible to complete a system of ($p < n$) linearly independent vectors $V_1, V_2, \ldots, V_p$ by $n - p$ vectors $V_{p+1}, V_{p+2}, \ldots, V_n$ such that the set forms a basis of E_n.*

The vector subspace E_p generated by these p vectors $V_1, V_2, \ldots, V_p$ is not identical to the space E_n, since this would mean that E_n is p-dimensional (Theorem (I-32)). The set $E_n - E_p$ is therefore not empty and we can consider a vector U in it. The set $V_1, V_2, \ldots, V_p, U$ is linearly independent since if it were not U would be a linear combination of $V_1, V_2, \ldots, V_p$ which is impossible since U does not belong to E_n.

We have completed the p vectors $V_1, V_2, \ldots, V_p$ by a vector U such that the set of $p + 1$ vectors $V_1, V_2, \ldots, V_p, U$ is linearly independent. Setting

$U = V_{p+1}$ and repeating the process, we obtain n linearly independent vectors, that is, a basis of E_n.

A SPECIAL VECTOR SPACE

So far we have defined vector spaces only by their properties, without actually constructing them.

We now construct a special but very important type of vector space. We first define what is meant by a "vector" and then give the rules of operation on the vectors under consideration. We than show that the sets established in this way form a vector space.

(I-36) *We call vector and denote by* $(a_1, a_2, \ldots, a_n)$ *any set of* n *scalars* $a_1, a_2, \ldots, a_n$, *taken from a field* K *and considered as a whole, in the given order. These elements are called the components of the vector.*

In most cases the field K will be the field of real numbers, sometimes the field of complex numbers. All the results that will be established, with a few exceptions, will be valid in either case.

The sets $\mathcal{E}$ under consideration consist therefore of the groups of n scalars considered as a whole in a well-determined order. To form a vector space with these elements, we define a neutral element, an addition operation and an operation of multiplication by scalars.

DEFINITION OF A NEUTRAL ELEMENT

By definition, the neutral element θ *in the set* $\mathcal{E}$ *is the element in which all components are zero:*

$$\theta = (0, 0, \ldots, 0) .$$

DEFINITION OF ADDITION

Given two elements $(a_1, a_2, ..., a_n)$ *and* $(b_1, b_2, ..., b_n)$ *in* $\mathcal{E}$, *the element* $(a_1 + b_1, a_2 + b_2, ..., a_n + b_n)$ *is said to be the sum of these two elements. The components of the sum element of two elements of* $\mathcal{E}$ *are the sums of the components of same rank of these elements:*

$$(a_1, a_2, ..., a_n) + (b_1, b_2. ..., b_n) = (a_1 + b_1, a_2 + b_2, ..., a_n + b_n). \tag{I-37}$$

DEFINITION OF MULTIPLICATION BY A SCALAR

Given an element $(a_1, a_2, ..., a_n)$ *in* $\mathcal{E}$, *the product of this element by a scalar* α *is the element* $(\alpha a_1, \alpha a_2, ..., \alpha a_n)$ *in* $\mathcal{E}$:

$$\alpha(a_1, a_2, ..., a_n) = (\alpha a_1, \alpha a_2, ..., \alpha a_n) . \tag{I-38}$$

With these definitions, it is easy to see that the set $\mathcal{E}$ forms a vector space E, since all the properties imposed on a vector space are satisfied by the structure we have just defined.

We now consider the following vectors in this space:

$$\begin{aligned} e_1 &= (1, 0, 0, ..., 0) , \\ e_2 &= (0, 1, 0, ..., 0) , \\ &\vdots \\ e_n &= (0, 0, 0, ..., 1) . \end{aligned}$$

The set of linear combinations of these vectors forms a vector space (I-27) and in general, applying the rules of operation defined above, we have:

$$\begin{aligned} V &= v_1 e_1 + v_2 e_2 + \cdots + v_n e_n \\ &= (v_1, v_2, ..., v_n) . \end{aligned}$$

It follows that the vectors e_i are linearly independent since V can be the zero vector θ only if all the v_i are zero.

On the other hand, it follows from the very form of the vector V that any vector of E can be expressed as a linear combination of the vectors e_i, since

$$(a_1, a_2, \ldots, a_n) = a_1 e_1 + a_2 e_2 + \cdots + a_n e_n.$$

Since there are n vectors e_i, the vector space E must be n-dimensional. Whence the theorem:

(I-39) *The vector space formed by the sets of n scalars $(a_1, a_2, \ldots, a_n)$ considered as a whole and in this order, is n-dimensional.* It is denoted E_n.

VECTOR SUBSPACES OF E_n

The first p vectors $e_i (p < n)$, generate a vector subspace of E_n whose vectors V' are of form

$$V' = v_1 e_1 + v_2 e_2 + \cdots + v_p e_p = (v_1, v_2, \ldots, v_p).$$

But we could just as well write

$$V' = v_1 e_1 + v_2 e_2 + \cdots + v_p e_p + 0\, e_{p+1} + \cdots + 0\, e_n$$
$$= (v_1, v_2, \ldots, v_p, 0, 0, \ldots, 0).$$

In other words, the vectors of E_n of form

$$(a_1, a_2, \ldots, a_p, 0, 0, \ldots, 0)$$

form a p-dimensional vector subspace of E_n.

ISOMORPHISM OF VECTOR SPACES

The n-dimensional vector space we have just defined has the important and very useful property of being isomorphic to all vector spaces of dimension n.

The concept of isomorphism can be defined as follows.

We have defined a vector space as a set of elements with certain properties, with which are associated two operations + and ×, the first being associated to all pairs of *E*, the second to all pairs whose first element belongs to the field *K* and the second to the space *E*. We say that the + operation is associated with the pairs of $E \times E$, while the × operation is associated with the pairs of $K \times E$(¹).

It is possible to conceive of another vector space E' characterized by other elements and by two other operations *R* and *T* respectively defined over every pair of $E' \times E'$ and over every pair of $K \times E'$.

In both cases, the elements and operations under consideration will always have the properties we have stated and described earlier (properties I-1 to I-21).

If it is possible to establish a one-to-one correspondence between the elements of the first space and those of the second and if this correspondence extends to the combinations of pairs by the defined operations, that is, if

$$\begin{matrix} V_1 \to V_1' \\ V_2 \to V_2' \end{matrix} \quad \text{and} \quad \begin{matrix} V_1 + V_2 \to V_1' \, R V_2' \\ a \times V \to a T V', \end{matrix}$$

then the two vector spaces under consideration are said to be isomorphic.

Let us therefore consider an n-dimensional vector space E_n defined over the field *K* with the operations + and × and the vector space consisting of the elements

(¹) The symbol $K \times E$ denotes the set of pairs whose first element is in *K* and the second in *E*.

$(a_1, a_2, \ldots, a_n)$ with the two operations defined a little earlier (the a_i are taken from the field K).

Consider in E_n a basis consisting of n linearly independent vectors $A_1, A_2, \ldots, A_n$. By Theorem (I-33) every vector V of E_n can be uniquely expressed as a linear combination of the A_i:

$$V = v_1 A_1 + v_2 A_2 + \cdots + v_n A_n .$$

Let $V \in E_n$ correspond to $V' \in E_n'$ such that:

$$V' = (v_1, v_2, \ldots, v_n).$$

The correspondence is one-to-one since the representation of V is unique. If on the other hand we consider two vectors V and U in E_n:

$$V = v_1 A_1 + v_2 A_2 + \cdots + v_n A_n ,$$
$$U = u_1 A_1 + u_2 A_2 + \cdots + u_n A_n ,$$

and the corresponding vectors in E_n' :

$$V' = (v_1, v_2, \ldots, v_n) ,$$
$$U' = (u_1, u_2, \ldots, u_n) ,$$

we have

$$V + U = (v_1 + u_1) A_1 + (v_2 + u_2) A_2 + \cdots + (v_n + u_n) A_n$$

and also

$$V' + U' = (v_1 + u_1, v_2 + u_2, \ldots, v_n + u_n) .$$

The correspondence between V and V' on the one hand, between U and U' on the other, implies therefore the correspondence between $V + U$ and $V' + U'$.

N.B. We have used our simplified notation for the operations, but it is of course understood that we should have **written** $V \oplus U$ and $V' \oplus U'$.

A similar result can be deduced for the second operation.

(I-40) *The space* E_n' *is therefore isomorphic to the space* E_n.

This means that the vector spaces are identical since they only differ in the name or the form of their elements and operations.

CHAPTER II

LINEAR MAPPINGS. MATRICES. GENERAL PROPERTIES

The theory of linear mappings and the theory of matrices are developed in parallel in this chapter, and we shall see at the end of the chapter that the linear mappings of a space E_n into a space E_m and the matrices with m rows and n columns are isomorphic.

Consider two vector spaces E_n and E_m with dimension n and m respectively, and suppose a correspondence has been defined which associates with any vector X of E_n a vector Y of E_m.

This correspondence is represented by the following notation:

$$Y = f(X), \qquad X \in E_n, \qquad Y \in E_m.$$

f is said to be a mapping of the space E_n into the space E_m and Y is the image of the vector X under the mapping f.

The space E_m can be the space E_n itself, that is, the space E_n can be mapped into itself.

If the mapping f has the two following properties:

$$f(X_1 + X_2) = f(X_1) + f(X_2), \tag{II-1}$$

$$f(\alpha \times X) = \alpha \times f(X) \tag{II-2}$$

it is said to be a linear mapping.

We therefore see that a linear mapping is merely a homomorphism in the context of vector spaces.

It is essential to note here that the equalities (II-1) and (II-2) are actually meaningful, which is not evident a priori.

To see that they are meaningful, note that in the left-hand side X_1 and X_2 belong to E_n, and therefore so does $X_1 + X_2$. Consequently $f(X_1 + X_2)$, the image of a vector in E_n, is a vector in E_m. Now in the right-hand side, $f(X_1)$ and $f(X_2)$ are both in E_m, and therefore so is their sum. The relation (II-1) is therefore meaningful. Similarly, the equality (II-2) is also meaningful.

The zero mapping is the mapping f under which the images of all the vectors $X \in E_n$ are the vector $\theta \in E_m$:

(II-3) $$f(X) = \theta \quad \text{for any} \quad X \in E_n .$$

We immediately see that it is a linear mapping.

Two linear mappings f and g of E_n into E_m are identical if

$$f(X) = g(X)$$

for all the vectors X in E_n. We then write

$$f \equiv g .$$

It is clearly not always easy to verify such an equality for all the vectors in E_n, but the following theorem makes this verification much easier.

(II-4) *Two linear mappings of E_n into E_m are identical if and only if the relation $f(X) = g(X)$ is satisfied by n linearly independent vectors $A_1, A_2, \ldots, A_n$ of E_n.*

The vectors A_i form a basis of E_n and thus

$$f(X) = x_1 f(A_1) + x_2 f(A_2) + \cdots + x_n f(A_n),$$
$$g(X) = x_1 g(A_1) + x_2 g(A_2) + \cdots + x_n g(A_n)$$

for any $X \in E_n$.

Since $f(A_i) = g(A_i)$ for all i, we immediately deduce:

$$f(X) = g(X).$$

REPRESENTATION OF A LINEAR MAPPING

The first problem encountered in the study of linear mappings is to find a way of representing them.

To do this, we consider a basis of E_n and a basis of E_m. The first will consist of n linearly independent vectors $A_1, A_2, \ldots, A_n$, the second of m linearly independent vectors $B_1, B_2, \ldots, B_m$.

Given a vector $X \in E_n$, we know it can be expressed as follows:

$$X = x_1 A_1 + x_2 A_2 + \cdots + x_n A_n .$$

Using relations (II-1) and (II-2), we can then write

$$f(X) = x_1 f(A_1) + x_2 f(A_2) + \cdots + x_n f(A_n)$$

which gives us the very important result:

(II-5) *The images of all the vectors of E_n under a linear mapping of E_n into E_m form the vector subspace generated in E_m by the images of the vectors of a basis of E_n.*

A linear mapping is therefore completely determined as soon as we know the images

$$f(A_1), \quad f(A_2), \ldots, f(A_n).$$

We now express these images in terms of the basis vectors $B_1, B_2, \ldots, B_m$.

All the vectors $f(A_j)$ can be expressed in terms of the vectors B_i and we have:

$$\begin{aligned} f(A_1) &= a_1^1 B_1 + a_2^1 B_2 + \cdots + a_m^1 B_m , \\ f(A_2) &= a_1^2 B_1 + a_2^2 B_2 + \cdots + a_m^2 B_m , \\ &\cdot \\ &\cdot \\ &\cdot \\ f(A_n) &= a_1^n B_1 + a_2^n B_2 + \cdots + a_m^n B_m . \end{aligned}$$

We can represent these relations by means of the following rectangular array:

$$\begin{bmatrix} a_1^1 & a_1^2 & \ldots & a_1^j & \ldots & a_1^n \\ a_2^1 & a_2^2 & \ldots & a_2^j & \ldots & a_2^n \\ \cdot & \cdot & \cdot & \cdot & \cdot & \cdot \\ a_i^1 & a_i^2 & \ldots & a_i^j & \ldots & a_i^n \\ \cdot & \cdot & \cdot & \cdot & \cdot & \cdot \\ a_m^1 & a_m^2 & \ldots & a_m^j & \ldots & a_m^n \end{bmatrix} = [A] = [a_i^j] .$$

Let us examine this array. We see that it contains the coefficients of the vectors B_i in the expressions for the vectors $f(A_j)$, but the coefficients which stood in a row in the system of relations given above now form a column, and conversely.

This rotation was performed in order to give the indices a presentation compatible with standard usage. The vector interpretation of the array is as follows: Each column j represents the sequence of the components of the vector $f(A_j)$ on the various basis vectors B_i and an element a_i^j is the component of the vector $f(A_j)$ on the vector B_i.

Such an array is called a matrix with m rows and n columns, or an m × n matrix, the latter notation being used for simplicity.

We immediately see that the matrix $[A]$ is not only connected with the vectorial nature of the linear mapping but also with the bases used in the spaces E_n and E_m; if other bases were used the matrix would be changed in a manner we shall study in detail in the next chapter.

We then have the following result:

If the vector space E_n has a basis $A_1, A_2, \ldots, A_n$ and the vector space E_m has a basis $B_1, B_2, \ldots, B_m$ there is a one-to-one correspondence between the linear mappings of E_n into E_m and the matrices with m rows and n columns, or $m \times n$ matrices.

Only one matrix can correspond to a mapping $f(X)$. We have seen earlier that the mapping f is completely defined by the images $f(A_1), f(A_2), \ldots, f(A_n)$. In E_m, these images can only be represented in one way as linear combinations of the vectors B_i. It follows that the coefficients a_i^k are uniquely determined.

Conversely, consider the matrix $[A]$ whose elements are the a_i^k. Given a column k, we can always consider the vectors Y_k defined as follows:

$$Y_k = a_1^k B_1 + a_2^k B_2 + \cdots + a_n^k B_n$$

and we can set

$$Y_k = f(A_k), \qquad k = 1, 2, \ldots, n,$$

setting up a correspondence between the vectors Y_k and the vectors $A_1, A_2, \ldots, A_n$. We can immediately verify that this correspondence defines the linear mapping with matrix $[A]$.

Suppose the matrix $[A]$ were to correspond to another linear mapping g. We would have, for the vectors A_i:

$$f(A_i) = g(A_i),$$

and consequently, by Theorem II-4, we would have $f(X) = g(X)$ for any $X \in E_n$, and the mappings f and g would be identical.

In other words, a unique matrix $[A]$ corresponds to the linear mapping $f(X)$ and a unique linear mapping $f(X)$ corresponds to the matrix $[A]$.

Consequently, two matrices $[A]$ and $[B]$ are identical if

$$[A]\,X = [B]\,X \quad \text{for any} \quad X \in E_n$$

which requires that their elements be equal one by one, and we then write $[A] = [B]$.

N.B. Note that the matrix corresponding to the zero linear mapping is

$$[0] = \begin{bmatrix} 0 & 0 & \cdots & 0 \\ \vdots & \vdots & & \\ \vdots & \vdots & & \\ 0 & 0 & \cdots & 0 \end{bmatrix}$$

in which all elements are zero.

OPERATIONS DEFINED ON LINEAR MAPPINGS AND MATRICES

We now define two operations on linear mappings. We denote the first by S and call it "sum": to two linear mappings f and g of E_n into E_m it associates a mapping k of E_n into E_m obtained by setting up a correspondence between each $X \in E_n$ and the vector $Y \in E_m$ obtained as follows:

$$Y = f(X) + g(X) = (f \, S \, g)(X) = k(X) .$$

The second operation, denoted 0, associates with a scalar $\alpha \in K$ and with a linear mapping f of E_n into E_m, a mapping p of E_n into E_m obtained by setting up a correspondence between each vector $X \in E_n$ and the vector $Y \in E_m$ obtained as follows

$$Y = \alpha \otimes f(X) = (\alpha \, 0 f)(X) = p(X) .$$

It will readily be verified that both operations have the respective properties (I-1) to (I-5) and (I-17) to (I-21).

For example, let us show that they are closed. We have indeed

$$\begin{aligned} k(X_1 + X_2) &= (f \, S \, g)(X_1 + X_2) = f(X_1 + X_2) + g(X_1 + X_2) \\ &= f(X_1) + f(X_2) + g(X_1) + g(X_2) \\ &= f(X_1) + g(X_1) + f(X_2) + g(X_2) = (f S g)(X_1) + (f S g)(X_2) \\ &= k(X_1) + k(X_2) \end{aligned}$$

and

$$\begin{aligned} k(cX) &= (f \, S \, g)(cX) = f(cX) + g(cX) = cf(X) + cg(X) \\ &= c[f(X) + g(X)] = c(f \, S \, g)(X) = ck(X) . \end{aligned}$$

The mapping k is therefore linear and the operation S is closed.

Similarly

$$\begin{aligned} p(X_1 + X_2) &= (\alpha \, 0 f)(X_1 + X_2) = \alpha f(X_1 + X_2) \\ &= \alpha f(X_1) + \alpha f(X_2) = (\alpha \, 0 f)(X_1) + (\alpha \, 0 f)(X_2) \\ &= p(X_1) + p(X_2) \end{aligned}$$

and

$$\begin{aligned} p(cX) &= (\alpha \, 0 f)(cX) = \alpha f(cX) = \alpha c f(X) = c\alpha f(X) \\ &= c(\alpha \, 0 f)(X) = cp(X) . \end{aligned}$$

The operation 0 is therefore also closed.

It is easy to prove the other properties.

It follows from these considerations that the set of linear mappings of a space E_n into a space E_m forms a vector space whose dimension will be determined at a later stage.

For the time being we will be satisfied to note that the operations S and 0 defined above are closed, so that, since the mappings f and g are linear, so are the mappings $f \; S \; g$ and $\alpha \; 0 \; f$.

Consequently, two $m \times n$ matrices will correspond to these mappings, and if we denote by $[A]$ and $[B]$ the matrices corresponding to f and to g, respectively, we say that the matrix corresponding to $f \; S \; g$ is the sum $[S]$ of the matrices $[A]$ and $[B]$, and that the matrix corresponding to $\alpha \; 0 \; f$ is the product of the matrix $[A]$ by the scalar α.

In order to determine the elements of these two matrices, we note that

$$(f \, S \, g) \, A_k = f(A_k) + g(A_k) = \sum a_i^k \, B_i + \sum b_i^k \, B_i = \sum (a_i^k + b_i) \, B_i = \sum s_i^k \, B_i \, .$$

Whence the following result:

(II-7) *The sum of two matrices $[A]_{m \times n}$ and $[B]_{m \times n}$ is a matrix $[S]_{m \times n}$ whose elements s_i^k are the sums of the corresponding elements a_i^k and b_i^k:*

$$[a_i^k] + [b_i^k] = [a_i^k + b_i^k].$$

The following result concerning the product of a matrix $[A]$ by a scalar α can be proved in a similar fashion:

(II-8) *The product of a matrix* $[A]_{m\times n}$ *by a scalar* α *is an* $m \times n$ *matrix whose elements* t_i^k *are the products by* α *of the elements* a_i^k:

$$\alpha[a_i^k] = [\alpha a_i^k].$$

It can readily be verified--we suggest that the reader do this as an exercise--that the operations of sum and product by a scalar defined on the matrices with m rows and n columns satisfy exactly the same properties as the analogous operations defined on the linear mappings of a space E_n into a space E_m. It follows that the vector spaces which form the linear mappings of E_n into E_m and the rectangular matrices with m rows and n columns are isomorphic and in particular that they have the same dimension.

This dimension is readily determined for the matrix form. All $m \times n$ matrices

$$[A] = \begin{bmatrix} a_1^1 & a_1^2 & \dots & a_1^n \\ a_1^2 & a_2^2 & \dots & a_2^n \\ \vdots & \vdots & & \vdots \\ a_m^1 & a_m^2 & \dots & a_m^n \end{bmatrix}$$

can be set in the following form:

$$[A] = \sum_i \sum_j a_i^j \, [E_i^j]$$

where

$$E_i^j = \begin{bmatrix} & \vdots^{\,j} & \\ \cdots & 1 & \cdots \\ & \vdots & \end{bmatrix} i$$

All elements of the matrix $[E_i^j]_{m \times n}$ are equal to zero except the element at the intersection of the ith row and the jth column, which is equal to 1.

The matrices $[E_i^j]$ are linearly independent since the only way of obtaining the matrix [0] from a linear combination of the matrices $[E_i^j]$ is to take all the coefficients a_i^j equal to zero. Thus, by Theorem I-32, the dimension of the vector space of the matrices $[A]$ with m rows and n columns is equal to the number of matrices $[E_i^j]$, that is, mn.

(II-9) *The vector space of all linear mappings of the space E_n into the space E_m, or of all $m \times n$ matrices, is of dimension mn.*

COMPOSITION OF LINEAR MAPPINGS. PRODUCTS OF MATRICES

We now introduce a new concept which we did not meet when dealing with vector spaces, that of the product of two elements of same nature; so far we have only considered their sum.

Let E_n, E_m, E_p be vector spaces with dimension n, m and p, respectively. Consider a linear mapping f of E_n into E_m and a linear mapping g of E_m into E_p.

To every vector V of E_n there corresponds a vector U of E_m defined by the relation

$$U = f(V), \quad U \in E_m .$$

To every vector U defined in this way there corresponds a vector T of E_p defined by the relation

$$T = g(U), \quad T \in E_p .$$

Hence, to every vector V of E_n there corresponds a vector T of E_p and we are now going to study the properties of the corresponding mapping, called the product of the two mappings f and g.

We can write:

$$T = g[f(V)] = gf(V) .$$

And we have

$$gf(V_1 + V_2) = g[f(V_1 + V_2)] = g[f(V_1) + f(V_2)] .$$

In this relation, $f(V_1)$ and $f(V_2)$ are two vectors in E_m. Now g is a linear mapping of E_m into E_p; therefore

$$g[f(V_1) + f(V_2)] = g[f(V_1)] + g[f(V_2)] = gf(V_1) + gf(V_2) ,$$

that is

$$gf(V_1 + V_2) = gf(V_1) + gf(V_2) .$$

In exactly the same way, we see that

$$gf(\alpha V) = \alpha gf(V) .$$

It follows that gf is a linear mapping of the vector space E_n into the vector space E_p. A rectangular matrix $[P]_{p\times n}$ will therefore correspond to it; we call this matrix the product of the matrices [A] and [B] which represent the mappings f and g respectively.

To define the elements of the matrix [P] we must look for the representation of the basis vectors $A_1, A_2, \ldots, A_n$ of E_n in terms of the basis vectors $C_1, C_2, \ldots, C_p$ of E_p. We start with the representation of the vectors A_k in terms of the basis vectors $B_1, B_2, \ldots, B_m$ of E_m, and the representation of the vectors B_j in terms of the vectors C_i of E_p.

Thus

$$f(A_k) = \sum_{j=1}^{m} a_j^k B_j \qquad k = 1, 2, \ldots, n ,$$

$$g(B_j) = \sum_{i=1}^{p} b_i^j C_i \qquad j = 1, 2, \ldots, m ,$$

and consequently

$$gf(A_k) = g[f(A_k)] = g\left[\sum_{j=1}^{m} a_j^k B_j\right] = \sum_{j=1}^{m} a_j^k g(B_j)$$

$$= \sum_{j=1}^{m} a_j^k \left(\sum_{i=1}^{p} b_i^j C_i\right) = \sum_{i} C_i \left(\sum_{j=1}^{m} b_i^j a_j^k\right) .$$

The component of the vector $gf(A_k)$ on the vector C_i is therefore equal to

$$\sum_{j=1}^{m} b_i^j a_j^k .$$

We have thus, denoting by p_i^k the general element of the matrix corresponding to the mapping gf:

$$p_i^k = \sum_j b_i^j a_j^k$$

and we write

$$[P] = [B][A],$$

since we have denoted by $[P]$ the product matrix of the matrices $[B]$ and $[A]$ *in this order*.

Note that the matrix $[P]$ has the same number p of rows as the first matrix of the product, $[B]$, and the same number n of columns as the second matrix $[A]$, which can be written as follows:

$$[P]_{p\times n} = [B]_{p\times m}[A]_{m\times n}.$$

PROPERTIES OF PRODUCTS OF LINEAR MAPPINGS AND PRODUCTS OF MATRICES

We saw earlier that the product of two linear mappings is a linear mapping.

We cannot say, however, that this product operation is closed since it sets up a correspondence between elements which are not exactly of the same nature: a mapping of E_n into E_p with a mapping of E_n into E_m and with a mapping of E_m into E_p.

On the other hand, the product operation is closed in the usual sense of the word when it deals only with mappings of a space E_n into itself.

In all cases:

(II-10) *The product operation and the corresponding multiplication of matrices are associative and distributive with respect to the operation S defined earlier.*

This means that

$$hgf = h(gf) = hg(f)$$

and

$$h(fSg) = hfShg\,.$$

We prove this for the first operation. For any X in E_n we must have

$$h\{g[f(X)]\} = hgf(X) = (hg)f(X)\,.$$

The left-hand side represents the change of the vector X of E_n into the vector Y of E_m, that of the vector Y of E_m into the vector Z of E_p and that of the vector Z of E_p into the vector T of E_r. The middle member represents the change of the vector X of E_n into the vector T of E_r. The two members are therefore identical. An analogous result holds for the third form. In matrix form, we must show that

$$[C]\,[B]\,[A] = [C]\,([B]\,[A]) = ([C]\,[B])\,[A]\,.$$

Now this merely represents the following relations:

$$\sum_j \sum_l c_i^j\, b_j^l\, a_l^k = \sum_j c_i^j\, (\sum_l b_j^l\, a_l^k) = \sum_l\, (\sum_j c_i^j\, b_j^l)\, a_l^k\,.$$

We now show that the product operation is distributive only for the matrix form.

We must have

$$\sum_j c_i^j(a_j^k + b_j^k) = \sum_j c_i^j a_j^k + \sum_j c_i^j b_j^k .$$

This equality follows from the properties of scalars when we perform the operations in the left-hand side.

It is important to note that compositions of linear mappings and products of matrices are not in general commutative.

We have shown earlier that there is a one-to-one correspondence between the linear mappings of a space E_n into a space E_m and the rectangular $m \times n$ matrices. We now can complete this correspondence with the representation of the image vectors by means of matrices, thanks to the concept of vector product.

We have seen that the image of a vector X under the mapping f can be expressed as

$$f(X) = x_1 f(A_1) + x_2 f(A_2) + \cdots + x_n f(A_n) .$$

Let us replace $f(A_1), f(A_2), \ldots, f(A_n)$ by their expressions in terms of $B_1, B_2, \ldots, B_m$. We have

$$f(X) = x_1(a_1^1 B_1 + a_2^1 B_2 + \cdots + a_m^1 B_m) + x_2(a_1^2 B_1 + a_2^2 B_2 + \cdots + a_m^2 B_m)$$
$$+ \cdots + x_n(a_1^n B_1 + a_2^n B_2 + \cdots + a_m^n B_m) .$$

Hence, if we denote by y_i the ith component of $f(X)$ in the space E_m, that is, the coefficient of B_i, then

$$y_i = \sum_{j=1}^{n} a_i^j x_j .$$

In other words, the vector $Y = f(X)$ can be represented as the product of two matrices: the matrix $[A]$ by the matrix $[X]$ with one column and n rows, and we can write

$$Y = f(X) = [A]\,[X].$$

In view of this representation, we call the vectors of the vector spaces under consideration *column vectors*, except in certain cases in which we shall adopt a special notation.

We can therefore interpret the various matrices we shall meet as follows:(1)

An $m \times n$ matrix represents a linear mapping of a space E_n into a space E_m.

An $n \times n$ matrix represent a mapping of the space E_n into itself.

An $n \times 1$ matrix represents a vector of the space E_n. Such a matrix $[X]$ is usually denoted without brackets, X.

A 1×1 matrix represents a scalar.

We shall see later on how to interpret $1 \times n$ matrices.

---

(1) There are other interpretations of the concept of matrix. Some will arise in the course of this book.

RANGE OF LINEAR MAPPINGS. RANK OF MATRICES

We have defined a linear mapping of a space E_n into a space E_m as a relation between the vectors of E_n and the vectors of E_m.

(II-11) *The set of vectors Y in E_m which are images of the vectors X in E_n under the mapping f is called the range of the linear mapping f. This range is a vector subspace of E_m* (in view of Theorem I-27). *The dimension of this vector subspace D is at most equal to the smallest of the numbers n and m.*

The vectors Y in the range D belong, by definition, to the space E_m so that $D \subset E_m$ and has dimension at most equal to m.

If n is greater than or equal to m, the theorem is proved. Suppose then that n is smaller than m, so that we must show that the dimension of D is at most equal to n. To do this, we show that it is possible, given p vectors of $D(p > n), Y_1, Y_2, \ldots, Y_p$ to find p scalars $a_1, a_2, \ldots, a_p$ not all zero, such that the linear combination

$$a_1 Y_1 + a_2 Y_2 + \cdots + a_p Y_p$$

is the zero vector θ, which will show that the vector space D has dimension at most n.

Since the Y_i belong to D, we have

$$Y_i = f(X_i)$$

where X_i is a vector in E_n; consequently

$$\alpha_1 Y_1 + \alpha_2 Y_2 + \cdots + \alpha_p Y_p = \alpha_1 f(X_1) + \alpha_2 f(X_2) + \cdots + \alpha_p f(X_p)$$
$$= f(\alpha_1 X_1 + \alpha_2 X_2 + \cdots + \alpha_p X_p)$$
$$X_1, X_2, \ldots, X_p \in E_n .$$

Since the vectors X_i belong to an n-dimensional space, it is always possible to find p scalars which are not all zero such that

$$\alpha_1 X_1 + \alpha_2 X_2 + \cdots + \alpha_p X_p = \theta$$

and we get

$$\alpha_1 Y_1 + \alpha_2 Y_2 + \cdots + \alpha_p Y_p = f(\theta) = \theta .$$

The following theorem throws even more light on these concepts:

(II-12) *The dimension of the range D is equal to the maximum number of linearly independent column vectors of the matrix* $[A]$. *This number r is called the rank of the matrix* $[A]$.

We have seen earlier that all the vectors of D can be expressed in terms of the column vectors of $[A]$, namely $f(A_1), f(A_2), \ldots, f(A_n)$ or $[A]\,A_1, [A]\,A_2, \ldots, [A]\,A_n$.

Suppose there are r linearly independent vectors $[A]\,A_i$ and that they are the first ones. It follows that the $n - r$ other vectors $[A]\,A_i$ can be expressed in terms of the r first ones, so that the vectors of D can be expressed in terms of these r vectors $[A]\,A_i$. By Theorem (I-32), the dimension of D is therefore r.

TRANSPOSE LINEAR MAPPINGS. TRANSPOSE MATRICES

The results we are now going to prove are valid for vector spaces and matrices defined over the field K of real numbers. Some slight changes would be necessary if we considered the field of complex numbers or more general fields.

We have given an interpretation of the columns of an $m \times n$ matrix; these are the components, on the vectors of the basis in E_m, of the images of the vectors of the basis in E_n, and we have seen that the linear mapping with matrix $[A]$ transforms a column vector of E_n into a column vector of E_m by premultiplication.

What interpretation can we give of the rows of this matrix? To find this, we transpose the matrix $[A]$, that is, we replace it by a matrix $[A]'$ the columns of which will be the rows of $[A]$ and conversely.

The matrix $[A]'$ contains n rows and m columns. It therefore represents a linear mapping of the space E_m into the space E_n. *This mapping f' is said to be the transpose mapping of the mapping f.*

We are now going to establish several very important results concerning transpose matrices and transpose linear mappings. First of all:

(II-13) *The dimension of the range of the mapping f' with matrix $[A]'$ is the same as that of the range of the mapping f with matrix $[A]$.*

We shall prove this result for matrices, stating the theorem in a slightly different way. We can indeed note that the dimension of the range of f is the maximum number of linearly independent column vectors of $[A]$ and that the dimension of the range of f' is the maximum number of linearly independent row vectors of $[A]$. We therefore state the theorem as follows:

(II-14) *The maximum number r of linearly independent column vectors in an $m \times n$ matrix is equal to the maximum number ρ of linearly independent row vectors. This common value is called the rank of the matrix.*

Consider a matrix

$$\begin{bmatrix} a_1^1 & a_1^2 & \dots & a_1^n \\ a_2^1 & a_2^2 & \dots & a_2^n \\ \cdot & \cdot & \cdot & \cdot \\ a_m^1 & a_m^2 & \dots & a_m^n \end{bmatrix}$$

This matrix contains n column vectors A with m component
m row vectors L with n component

Consider the following linear relationship between p column vectors, where $p > \rho$:

$$\lambda_1 A_1 + \lambda_2 A_2 + \cdots + \lambda_p A_p = \theta .$$

We will show that it is always possible to find p scalars λ_i which are not all zero, such that this relation is satisfied. It will then follow that the p vectors A_i are not linearly independent.

If we consider coordinates, this vector equation is equivalent to the following m numerical equations:

$$(I) \begin{cases} \lambda_1 a_1^1 + \lambda_2 a_1^2 + \cdots + \lambda_p a_1^p = \Lambda_1 = 0 , \\ \lambda_1 a_2^1 + \lambda_2 a_2^2 + \cdots + \lambda_p a_2^p = \Lambda_2 = 0 , \\ \cdots \\ \lambda_1 a_k^1 + \lambda_2 a_k^2 + \cdots + \lambda_p a_k^p = \Lambda_k = 0 , \\ \cdots \\ \lambda_1 a_m^1 + \lambda_2 a_m^2 + \cdots + \lambda_p a_m^p = \Lambda_m = 0 . \end{cases}$$

Suppose the ρ first row vectors $L_1, L_2, \ldots, L_\rho$ are linearly independent. This means that the m-ρ row vectors $L_{\rho+1}, L_{\rho+2}, \ldots, L_m$ are linear combinations of these ρ first vectors, that is,

$$L_k = \mu_k^1 L_1 + \mu_k^2 L_2 + \cdots + \mu_k^\rho L_\rho$$

or, considering the components of L_k,

$$\begin{aligned}
a_k^1 &= \mu_k^1 a_1^1 + \mu_k^2 a_2^1 + \cdots + \mu_k^\rho a_\rho^1 , \\
a_k^2 &= \mu_k^1 a_1^2 + \mu_k^2 a_2^2 + \cdots + \mu_k^\rho a_\rho^2 , \\
&\cdots\cdots\cdots\cdots\cdots\cdots \\
a_k^n &= \mu_k^1 a_1^n + \mu_k^2 a_2^n + \cdots + \mu_k^\rho a_\rho^n .
\end{aligned}$$

Substituting these values in the equation

$$\lambda_1 a_k^1 + \lambda_2 a_k^2 + \cdots + \lambda_n a_k^n = 0$$

of system (I), we get

$$\begin{aligned}
\lambda_1(\mu_k^1 a_1^1 + \mu_k^2 a_2^1 + \cdots + \mu_k^\rho + a_\rho^1) + \lambda_2(\mu_k^1 a_1^2 + \mu_k^2 a_2^2 + \cdots + \mu_k^\rho a_\rho^2) \\
+ \cdots + \lambda_n (\mu_k^1 a_1^n + \mu_k^2 a_2^n + \cdots + \mu_k^\rho a_\rho^n) = 0 ,
\end{aligned}$$

or, regrouping the terms in $\mu_k^1, \mu_k^2, \ldots, \mu_k^\rho$

$$\begin{aligned}
\mu_k^1(\lambda_1 a_1^1 + \lambda_2 a_2^1 + \cdots + \lambda_n a_1^n) + \mu_k^2(\lambda_1 a_2^1 + \lambda_2 a_2^2 + \lambda_n a_2^n) \\
+ \cdots + \mu_k^\rho(\lambda_1 a_\rho^1 + \lambda_2 a_\rho^2 + \cdots + \lambda_n a_\rho^n) = 0,
\end{aligned}$$

that is,

$$\Lambda_k = \mu_k^1 \Lambda_1 + \mu_k^2 \Lambda_2 + \cdots + \mu_k^\rho \Lambda_\rho = 0 .$$

This means that the m-ρ last equations of system (I) can be deduced from the ρ first ones. It follows that system (I) is equivalent to the system (I') obtained by deleting the $m-\rho$ last equations in (I):

$$(\mathrm{I}') \quad \begin{cases} \Lambda_1 = 0 \\ \Lambda_2 = 0 \\ \vdots \\ \Lambda_\rho = 0 \end{cases}$$

Indeed, every solution of (I) is a solution of (I') and conversely every solution of (I') is a solution of (I) since when the first ρ quantities Λ_ρ are zero so are the $m-\rho$ last ones.

Now the system (I') is a system with p unknowns $\lambda_1, \lambda_2, \ldots, \lambda_p$ and ρ equations, where $p > \rho$. It follows from Theorem (I-7) that it admits a non-trivial solution and that there exist coefficients λ_i not all zero, such that

$$\lambda_1 A_1 + \lambda_2 A_2 + \cdots + \lambda_p A_p = \theta\,,$$

for any p greater than ρ. The maximum number of linearly independent vectors A_i is therefore at most equal to ρ, that is

$$r \leqslant \rho\,.$$

We can carry out a similar reasoning starting with the row vectors, so that we obtain the following result:

$$\rho \leqslant r\,,$$

and, combining the two inequalities, we finally get

$$\rho = r .$$

It is also possible to state this theorem in another form:

(II-15) *A matrix* $[A]$ *and its transpose* $[A]'$ *have the same rank r.*

Tranposition transforms the sum and product operations of matrices as follows:

$$[A + B]' = [A]' + [B]' ,$$
$$([A]\,[B])' = [B]'\,[A]' .$$

The first relation is obvious. To prove the second, we compute both sides separately, and verify that they are equal.

Let p_i^k be the general term of the matrix $[A]\,[B]$, and q_i^k the general term of the matrix $([A]\,[B])'$. Then

$$q_i^k = p_k^i = \sum_j a_k^j\, b_j^i .$$

Let $b_i^{k'}$ and $a_i^{k'}$ be the general terms of the matrices $[B]'$ and $[A]'$; the general term r_i^k of matrix $[B]'\,[A]'$ is

$$r_i^k = \sum_j b_i^{j'}\, a_j^{k'} = \sum_j b_j^i\, a_k^j .$$

And thus

$$q_i^k = r_i^k .$$

We have seen, on page 40, that the vectors X in should be considered as $n \times 1$ matrices. If we transpose a column vector X, we obtain a row vector X' with 1 row and n columns. The fundamental importance of row vectors will be apparent in Chapter VI.

If a matrix $[A]$ *is equal to its transpose* $[A']$, *it is said to be symmetric:*

$$[A'] = [A].$$

It follows immediately that a symmetric matrix is necessarily a square matrix, i.e. $n = m$, and is such that

$$a_i^j = a_j^i.$$

(II-16) *Given any rectangular matrix* $[A]_{m \times n}$, *the matrix* $[P] = [A'][A]$ *is symmetric with* n *rows and* n *columns.*

We have indeed:

$$[P'] = ([A'][A])' = [A'][A']' = [A'][A] = [P]$$

since the transpose $[A']'$ of the transpose matrix is clearly the matrix $[A]$ itself.

CHAPTER III

LINEAR MAPPINGS OF A SPACE E_n INTO ITSELF AND THE CORRESPONDING MATRICES. CHANGES OF BASIS IN E_n

So far, we have considered linear mappings of a space E_n into a space E_m. A very important special case is when we consider a mapping of the space E_n into itself, that is, when to a vector X in E_n there corresponds another vector Y in E_n.

All the general properties given in Chapter II for linear mappings of a space E_n into a space E_m and for rectangular matrices with m rows and n columns also hold for linear mappings of the space E_n into itself. We need only set $m = n$ everywhere. Note however that the basis B_1, B_2, ..., B_m of E_m is replaced by the basis A_1, A_2, ..., A_n of E_n that was previously considered.

Moreover, linear mappings of E_n into itself and square matrices with n row and n columns have some very interesting properties which will be examined in this chapter.

We have seen earlier that the range D of a linear mapping of E_n into E_m is a subspace of E_m. In the case of a mapping of E_n into itself it is therefore either a subspace of E_n or all of E_n.

If the range D of the linear mapping f of E_n into itself is the whole space E_n, the mapping f is said to be non-singular and a mapping of E_n onto itself. The matrix of the mapping f is also said to be non-singular.

If D is a vector subspace of E_n, the mapping f is

said to be singular, and so is its matrix.

If the linear mapping f is non-singular, this means that every vector $Y \in E_n$ is the image of another vector $X \in E_n$.

There are several criteria allowing us to determine whether a linear mapping of E_n into itself is singular or not. We now examine the most important ones.

(III-1) *A linear mapping f is non-singular if and only if the vectors $f(A_1)$, $f(A_2)$, ..., $f(A_n)$ are linearly independent.*

The condition is necessary. If the mapping f is non-singular, the range of f is all of E_n, and is therefore an n-dimensional vector space. But the vectors Y of D are all linear combinations of the vectors $f(A_i)$. Suppose then that there are only p linearly independent vectors $f(A_i)$ $(p<n)$: it would follow that the dimension of D would be p, which contradicts the hypothesis that D is n-dimensional.

The condition is sufficient. If the vectors $f(A_i)$ are linearly independent, it follows from Theorem (I-32) that the vectors of D, which are all linear combinations of the $f(A_i)$, form an n-dimensional space, that is, the space E_n itself.

There is another proof for this converse. If the vectors $f(A_1)$, $f(A_2)$, ..., $f(A_n)$ which belong to E_n are linearly independent, this means that they form a basis of E_n. Consequently, any vector $U \in E_n$ can be expressed as follows:

$$\begin{aligned} U &= U_1 f(A_1) + U_2 f(A_2) + \cdots + U_n f(A_n) \\ &= f(U_1 A_1 + U_2 A_2 + \cdots + U_n A_n) = f(V) \end{aligned}$$

where

$$V = U_1 A_1 + U_2 A_2 + \cdots + U_n A_n .$$

In other words, every vector U in E_n is the image of another vector V in E_n.

This allows us to give another definition of a non-singular linear mapping:

(III-2) *A linear mapping f of the space E_n into itself is non-singular if every vector of E_n can be considered as the image under the mapping f of another vector V of E_n.*

We can also state the condition for non-singularity as follows:

(III-3) *A linear mapping is non-singular if and only if the only vector $X \in E_n$ for which*

$$f(X) = \theta$$

is the vector θ itself.

The condition is necessary. If the linear mapping f is non-singular, it follows that the vectors $f(A_i)$ are linearly independent. Now

$$f(X) = x_1 f(A_1) + x_2 f(A_2) + \cdots + x_n f(A_n) .$$

Therefore, in order that $f(X)$ be θ, all the x_i must be identically zero, that is, X must be θ.

The condition is sufficient. The vector $f(X)$ can be written in the form given above, and if $f(X)$ can be equal to θ only when $X = \theta$, this means that all the x_i must be identically zero, which means that the vectors $f(A_i)$ are linearly independent.

Theorem (III-3) has an immediate corollary:

(III-3a) *A non-singular matrix transforms linearly independent vectors into linearly independent vectors.*

Consider k linearly independent vectors V_1, V_2, ..., V_k. The k vectors $[A]V_1$, $[A]V_2$, ..., $[A]V_k$, where $[A]$ is a non-singular matrix, are linearly independent.

Indeed, if they were not, we would be able to find k scalars α_i, not all zero, such that:

$$\alpha_1[A]\,V_1 + \alpha_2[A]\,V_2 + \cdots + \alpha_k[A]\,V = \theta,$$

which can be expressed as

$$[A]\,(\alpha_1\,V_1 + \alpha_2\,V_2 + \cdots + \alpha_k\,V_k) = \theta\,,$$

which requires, since $[A]$ is non-singular:

$$\alpha_1\,V_1 + \alpha_2\,V_2 + \cdots + \alpha_k\,V_k = \theta,$$

and this latter relation is possible only if all the α_i are zero. It follows that the vectors $[A]V_i$ are indeed linearly independent.

We are now going to examine non-singular linear mappings.

NON-SINGULAR LINEAR MAPPINGS

We have seen earlier that every vector $Y \in E_n$ is the image of a vector $X \in E_n$ under a non-singular linear mapping and we can even state this result as follows:

Under a non-singular linear mapping, a vector $Y \in E_n$ is the image of a unique vector $X \in E_n$.

Suppose indeed that there were two vectors X and Z of which Y is the image under the linear mapping f. We would have:

$$f(X) = Y, \qquad f(Z) = Y,$$

that is

$$f(X) - f(Z) = \theta,$$

or

$$f(X) - f(Z) = f(X - Z) = \sum_i (x_i - z_i) f(A_i).$$

Now this linear combination of n linearly independent vectors $f(A_i)$ can be equal to the zero vector θ only if all its coefficients are zero, that is, if

$$x_i - z_i = 0 \quad \text{or} \quad x_i = z_i \quad \text{for } i = 1, 2, \ldots, n,$$

which implies

$$X = Z.$$

Since this proof is based on the independence of the vectors $f(A_i)$, this means that under a singular linear mapping, a vector Y in the range of f can be the image of several vectors of E_n.

We now define an important non-singular linear mapping of E_n onto itself, the identity mapping. This is the mapping under which every vector X corresponds to itself:

$$i(X) = X.$$

We immediately see that this mapping is linear and non-singular. The following matrix corresponds to it:

$$[I] = \begin{bmatrix} 1 & 0 & 0 & \ldots & 0 \\ 0 & 1 & 0 & \ldots & 0 \\ \cdot & \cdot & \cdot & \cdot & \cdot \\ 0 & 0 & 0 & \ldots & 1 \end{bmatrix}$$

which simply expresses the fact that the basis vectors A_1, A_2, ..., A_n are transformed into themselves.

INVERSE MAPPINGS

Consider a non-singular linear mapping f of E_n onto itself. To every vector $X \in E_n$ there corresponds an image $Z = f(X)$. And, by the previous theorem, the vector X corresponds to the vector Z. This correspondence of X with Z is a mapping g, that is

$$g(Z) = X \quad \text{and} \quad X = gf(X)\,.$$

If we consider two vectors Z_1 and Z_2, we can deduce from these two relations and from the fact that the mapping f is linear, that

$$\begin{aligned} g(Z_1 + Z_2) = g[f(X_1) + f(X_2)] &= gf(X_1 + X_2) = X_1 + X_2 \\ &= g(Z_1) + g(Z_2)\,. \end{aligned}$$

In the same way, if α is a scalar:

$$g(\alpha Z) = g[\alpha f(X)] = g[f(\alpha X)] = \alpha X = \alpha g(Z)\,.$$

(III-5) *It follows that the mapping g is linear and has a corresponding matrix $[B]_{n\times n}$.*

By analogy with the terminology used for the field of scalars, the mapping g is called the inverse mapping of the mapping f and the matrix $[B]$ is called the inverse of the matrix $[A]$. We have indeed:

$$g[f(X)] = X\,,$$

that is, in matrix notation,

$$[B]\,[A]\,X = [I]\,X$$

for any X.

It follows that

$$[B]\,[A] = [I]\,.$$

The matrix $[B]$ is usually denoted $[A]^{-1}$. We shall see on page 90 how its elements can be determined.

It is readily verified that the matrix $[A]$ is the inverse of the matrix $[A]^{-1}$. We have indeed:

$$[A]^{-1}\,[A] = [I]\,.$$

We can therefore write, applying the matrix $[A]$ to both sides of this equality:

$$[A]\,[A]^{-1}\,[A] = [A]\,[I] = [A] = [I]\,[A]\,.$$

It follows that

$$[A]\,[A]^{-1} = [I]$$

which shows that $[A]$ is the inverse of $[A]^{-1}$.

We also find the following result:

(III-6) *If the matrices $[A]$ and $[B]$ are non-singular, the matrix $[P] = [A][B]$ is also non-singular; it therefore admits an inverse which is defined by the following relation:*

$$[P]^{-1} = [B]^{-1}\,[A]^{-1}\,.$$

We first prove that $[P]$ is non-singular.

Since the matrix $[B]$ is non-singular, its column vectors $[B]A_1$, $[B]A_2$, ..., $[B]A_n$ are linearly independent; consequently, since the matrix $[A]$ is also non-singular, the vectors $[A][B]A_1$, $[A][B]A_2$, ..., $[A][B]A_n$ are linearly independent (Theorem (III-3a)). Now these vectors are the column vectors of the matrix $[P]$, which is therefore non-singular.

On the other hand

$$([B]^{-1}[A]^{-1})([A][B]) = [B]^{-1}[A]^{-1}[A][B] = \\ = [B]^{-1}[I][B] = [B]^{-1}[B] = [I].$$

Since the inverse of a matrix is unique, it follows that the matrix $[B]^{-1}[A]^{-1}$ is the inverse of the matrix $[A][B]$.

We now define another special linear mapping and its matrix.

The mapping of E_n into itself defined by $Y = kX$, where k is a non-zero scalar, is a linear mapping.

If we denote this mapping by $K(X)$, we can immediately verify that

$$K(X_1 + X_2) = k(X_1 + X_2) = kX_1 + kX_2 = K(X_1) + K(X_2),$$
$$K(\alpha X) = k\alpha X = \alpha k X = \alpha K(X).$$

Since the mapping K is linear, there is a matrix corresponding to it, the matrix $k[I]$,

$$k[I] = \begin{bmatrix} k & 0 & 0 & \dots & 0 \\ 0 & k & 0 & \dots & 0 \\ 0 & 0 & 0 & \dots & k \end{bmatrix}.$$

The corresponding mapping is non-singular and the above matrix is called a scalar matrix.[1]

The reader will no doubt have noticed the

---

(1) Scalar matrices will occur again on page 67.

fundamental role of non-singular linear mappings and their matrices, from the very beginning of this chapter. We now give a further example.

CHANGE OF BASIS IN E_n

We have seen earlier (page 12) that in an n-dimensional vector space every set of n linearly independent vectors $A_1, A_2, \ldots, A_n$ forms a basis of E_n, that is, that every vector $X \in E_n$ is a linear combination of the vectors A_i:

$$X_A = x_1^A A_1 + x_2^A A_2 + \cdots + x_n^A A_n .$$

where the scalars $x_1^A, x_2^A, \ldots, x_n^A$ are called the components of the vector X on the basis vectors $A_1, A_2, \ldots, A_n$.

If we choose another basis $B_1, B_2, \ldots, B_n$ in E_n, X can also be expressed as follows:

$$X_B = x_1^B B_1 + x_2^B B_2 + \cdots + x_n^B B_n$$

Given the components x_i^A of the vector X for the basis $A_1, A_2, \ldots, A_n$, we wish to find the components x_i^B of this same vector for the basis $B_1, B_2, \ldots, B_n$, or conversely. This is the change-of-basis problem.

Since all the vectors of E_n can be expressed in terms of the basis vectors B_i, it is sufficient to know the representation of these basis vectors B_i in terms of the vectors A_i, which is possible since the A_i form a basis of E_n.

We have thus

$$\text{(III-7)} \qquad \begin{cases} B_1 = b_1^1 A_1 + b_2^1 A_2 + \cdots + b_n^1 A_n , \\ B_2 = b_1^2 A_1 + b_2^2 A_2 + \cdots + b_n^2 A_n , \\ \text{-------------------------} \\ B_n = b_1^n A_1 + b_2^n A_2 + \cdots + b_n^n A_n . \end{cases}$$

To find the components x_i^A in terms of the components x_i^B it is sufficient to replace the B_j by their expressions above in the formula:

$$X = x_1^B B_1 + x_2^B B_2 + \cdots + x_n^B B_n .$$

We then find

$$X_B = x_1^B(b_1^1 A_1 + b_2^1 A_2 + \cdots + b_n^1 A_n) + \dot{x}_2(b_1^2 A_1 + b_2^2 A_2 + \cdots + b_n^2 A_n)$$
$$+ \cdots + x_n^B(b_1^n A_1 + b_2^n A_2 + \cdots + b_n^n A_n)$$

and, regrouping the terms in $A_1, A_2, \ldots, A_n$:

$$X_B = (x_1^B b_1^1 + x_2^B b_1^2 + \cdots + x_n^B b_1^n) A_1$$
$$+ (x_1^B b_2^1 + x_2^B b_2^2 + \cdots + x_n^B b_2^n) A_2 + \cdots$$
$$+ (x_1^B b_n^1 + x_2^B b_n^2 + \cdots + x_n^B b_n^n) A_n .$$

The coordinates of X in the basis $A_1, A_2, \ldots, A_n$ are therefore given by the following formulas in terms of the coordinates of X in the basis $B_1, B_2, \ldots, B_n$:

$$x_1^A = x_1^B b_1^1 + x_2^B b_1^2 + \cdots + x_n^B b_1^n ,$$
$$x_2^A = x_1^B b_2^1 + x_2^B b_2^2 + \cdots + x_n^B b_2^n ,$$
$$x_n^A = x_1^B b_n^1 + x_2^B b_n^2 + \cdots + x_n^B b_n^n .$$

This system of linear equations can be expressed as follows in matrix notation:

(III-8) $$\boxed{X_A = [B]\, X_B} .$$

The element b_i^k of the matrix $[B]$ represents the ith coordinate of the vector B_k in the basis $A_1, A_2, \ldots, A_n$.

Since the vectors $B_1, B_2, \ldots, B_n$ form a basis of E_n, they are linearly independent and the matrix $[B]$ is therefore non-singular. It therefore has an inverse matrix $[B]^{-1}$ which allows us to express X_B in terms of X_A, that is,

$$\boxed{X_B = [B]^{-1} X_A} . \qquad \text{(III-9)}$$

This formula allows us to express the coordinates of the vector X with respect to the basis $B_1, B_2, \ldots, B_n$ if we know its coordinates with respect to the basis $A_1, A_2, \ldots, A_n$.

We have just seen that the components of a vector X are not the same when the vector space E_n is defined by two different bases $A_1, A_2, \ldots, A_n$ and $B_1, B_2, \ldots, B_n$. In the same way, we expect the matrix representing a linear mapping to change when we go from one basis to another. We now examine this question.

TRANSFORMATION OF THE MATRICES OF LINEAR MAPPINGS

Consider a linear mapping with matrix $[F]$ from a space E_n with basis $A_1, A_2, \ldots, A_n$ into a space E_m with basis $C_1, C_2, \ldots, C_m$. In order to emphasize which basis is used, we write

$$Y_C = [F_{AC}] X_A .$$

If we use a basis $B_1, B_2, \ldots, B_n$ in E_n and a basis $D_1, D_2, \ldots, D_m$ in E_m, we can write similarly:

$$Y_D = [F_{BD}] X_B .$$

We will express $[F_{BD}]$ in terms of $[F_{AC}]$, using the formulas established earlier. We have indeed:

$$Y_C = [D] Y_D , \qquad X_A = [B] X_B ,$$

that is

$$[D]\, Y_D = [F_{AC}]\, [B]\, X_B \, .$$

Since the matrices $[B]$ and $[D]$ are non-singular, this implies

$$Y_D = [D]^{-1}\, [F_{AC}][B]\, X_B \, ,$$

that is,

(III-10) $$\boxed{[F_{BD}] = [D]^{-1}\, [F_{AC}]\, [B] \, .}$$

Note that the matrices in the right-hand side are written with respect to the initial bases A_1, A_2, ..., A_n for E_n and C_1, C_2, ..., C_m for E_m.

In the very common case of a mapping of E_n into itself, it is possible to write the two following relations:

$$Y_A = [F_A]\, X_A \, ,$$

$$Y_B = [F_B]\, X_B \, ,$$

and we get

(III-11) $$\boxed{[F_B] = [B]^{-1}\, [F_A]\, [B]} \, .$$

It is readily shown that the matrices $[F_B]$ and $[F_A]$ have the same rank. This is an immediate consequence of the following theorem:

(III-12) *The product of a square or rectangular matrix of rank r by a non-singular square matrix is a matrix of rank r.*

This is easy to prove by considering the corresponding linear mappings.

We therefore consider a matrix $[B]$ of rank r with m rows and n columns, and a non-singular matrix $[A]$ with n rows and n columns. The product $[A][B]$ is a matrix $[P]$ with m rows and n columns.

Now, the columns of the matrix $[P]$ are the vectors $[A][B]A_1$, $[A][B]A_2$, ..., $[A][B]A_n$, where A_1, A_2, ..., A_n are the basis vectors in the space E_n. But there are r linearly independent vectors among the vectors $[B]A_1$, $[B]A_2$, ..., $[B]A_n$ since the matrix $[B]$ has rank r. And, by Theorem (III-3a), these r column vectors $[B]A_i$ are transformed into r linearly independent vectors by the non-singular matrix $[A]$. The matrix P therefore contains precisely r linearly independent column vectors, and is thus of rank r.

If we had had to consider the product $[Q] = [B][A]$, where the non-singular matrix $[A]$ now has m rows and m columns, we would have considered the matrix $[Q'] = [A'][B']$ and we would have been brought back to the previous case since transposition does not affect the rank of a matrix.

We have repeatedly, in particular on page 29, emphasized the fact that there is a one-to-one relationship between the linear mappings of a space E_n into a space E_m and the rectangular $m \times n$ matrices, when the bases are fixed in the two spaces. This result is of course true for linear mappings of the space into itself and for square matrices.

If we change the bases in the vector spaces, the corresponding matrices will change also. It is therefore very important to know, given two $m \times n$ matrices, whether or not they represent the same linear mapping in different bases.

This leads us to introduce a new concept, that of equivalent matrices.

(III-13) *Two* $m \times n$ *matrices* $[A]$ *and* $[B]$ *are said to be equivalent if they satisfy the following relation:*

$$[B] = [P]\,[A]\,[R]$$

where $[P]$ *and* $[R]$ *are two non-singular* $m \times m$ *and* $n \times n$ *matrices, respectively.*

Under these conditions:

(III-14) *Two distinct* $m \times n$ *matrices represent the same linear mapping of a space* E_n *into a space* E_m *if and only if they are equivalent.*

The condition is necessary. If the two matrices $[F_{AC}]$ and $[F_{BD}]$ represent the same linear mapping of E_n into E_m, we have seen that they are connected by the following relation:

$$[F_{BD}] = [D]^{-1}\,[F_{AC}]\,[B]\,.$$

The matrices $[D]$, and consequently $[D]^{-1}$, and $[B]$ are non-singular since they represent changes of basis, so that the matrices $[F_{BD}]$ and $[F_{AC}]$ are indeed equivalent.

The condition is sufficient. Since the matrices $[F_{AC}]$ and $[F_{BD}]$ are equivalent, there exist two non-singular matrices $[P]$ and $[R]$ such that:

$$[F_{BD}] = [P]\,[F_{AC}]\,[R]\,.$$

Now the matrix $[F_{AC}]$ represents a certain linear mapping of the space E_n with basis $A_1, A_2, \ldots, A_n$ into a space E_m with basis $C_1, C_2, \ldots, C_m$. Thus, if we perform a change of basis in E_n defined by the matrix $[R]$ - which is possible since $[R]$ is non-singular - and in the same way a change of basis in E_m defined by the matrix $[P]^{-1}$ - which is possible since $[P]$, and therefore $[P]^{-1}$, is non-singular - we find that the representation of the linear mapping in this new basis is defined as follows:

$$[F_{RP}] = [[P]^{-1}]^{-1} [F_{AC}] [R] = [P] [F_{AC}] [R] = [F_{BD}] .$$

It follows that the matrices $[F_{AC}]$ and $[F_{BD}]$ must represent the same linear mapping.

It is not always easy to verify directly if two matrices are equivalent. The following theorem allows us to express the concept of equivalence in terms of vector independence

(III-15) *Two rectangular $m \times n$ matrices are equivalent if and only if they have the same rank.*

The condition is necessary. If the two matrices $[M]$ and $[N]$ are equivalent, they are connected by the following relation:

$$[N] = [P] [M] [R] ,$$

where $[P]$ and $[R]$ are two non-singular matrices. Now the product of a matrix of rank r by a non-singular matrix is a matrix of rank r (Theorem (III-12)). Hence the matrix $[P][M][R]$ is of rank r, like the matrix $[M]$.

The condition is sufficient. The matrices $[M]$ and $[N]$ have rank r and we suppose that the space E_n has a basis $A_1, A_2, \ldots, A_n$ and that the space E_m has a basis $B_1, B_2, \ldots, B_m$.

The matrix $[M]$ of rank r has r linearly independent column vectors $[M]A_1$, $[M]A_2$, ..., $[M]A_r$, and

$$[M] A_j = \sum_1^r \alpha_j^k [M] A_k \text{ for } j = r + 1, \ldots, n .$$

We define a change of basis in E_n by the following relations:

$$C_i = A_i \quad \text{for} \quad i = 1, 2, \ldots, r .$$

$$C_i = A_i - \sum_k^r \alpha_i^k A_k \quad \text{for} \quad i = r + 1, \ldots, n .$$

Since an arbitrary linear combination of the C_i is also a linear combination of the A_i, it follows that the C_i are linearly independent and form a basis of E_n.

Similarly we define a new basis in E_m by the following relations:

$$D_i = [M] A_i \quad \text{for} \quad i = 1, 2, \ldots, r ,$$

and complete the r vectors D_i by $m - r$ vectors D_j so as to form a complete basis of E_m. With these new bases, the matrix of the linear mapping will be of the following form:

$$[D]^{-1} [M_{AC}] [C] = \left[\begin{array}{ccc:ccc} \multicolumn{3}{c}{\overbrace{\hphantom{1 \quad 1 \quad 1}}^{r}} & \multicolumn{3}{c}{\overbrace{\hphantom{0 \quad 0 \quad 0}}^{n-r}} \\ 1 & & & & & \\ & 1 & & & & \\ & & 1 & & & \\ & & & 0 & & \\ & & & & 0 & \\ & & & & & 0 \end{array}\right]$$

that is, an $m \times n$ matrix in which all elements are zero except those on the main diagonal, r of which are equal to 1, while the other $n - r$ are zero. We can also verify that this matrix has rank r.

If we had started with the matrix $[N_{AC}]$ we would have obtained the same result with another change of basis, that is, we would have had

$$[D]^{-1}[M_{AC}][C] = [E]^{-1}[N_{AC}][F].$$

Multiplying both sides of this matrix relation by $[D]$ on the left and by $[C]$ on the right, we would get:

$$[M_{AC}] = [D][E]^{-1}[N_{AC}][F][C]^{-1} = [P][N_{AC}][R].$$

The matrices $[M_{AC}]$ and $[N_{AC}]$ are therefore equivalent.

Combining the two previous theorems, we get the following theorem:

(III-16) *Two district $m \times n$ matrices represent the same linear mapping of a space E_n into a space E_m if and only if they have the same rank.*

The previous concepts are slightly modified in the case of square $n \times n$ matrices.

(III-16a) *Two matrices $[A]_{n \times n}$ and $[B]_{n \times n}$ are said to be equivalent if they are connected by the following relation:*

$$[A] = [P][B][P]^{-1}$$

where $[P]$ is a non-singular $n \times n$ matrix.

Theorem (III-14) is valid for square matrices, but this is not the case for Theorems (III-15) and (III-16). This can be explained by the fact that for a linear mapping of the space E_n into itself the same basis is used to represent the domain vectors and the image vectors.

The theorems we have just proved allow us to appreciate the influence of a change of basis on linear mappings and to display the structural ties between matrices representing the same linear mapping.

Changes of basis will be used in the following

chapters to give linear mappings their simplest and most convenient representations. We will also examine the form of matrices associated with special and important linear mappings.

DIAGONAL AND SCALAR MATRICES

Consider a linear mapping characterized by the fact that each basis vector A_i is transformed into a vector B_i collinear with A_i:

$$[D]\,A_i = d_i\,A_i\,.$$

(III-17) *If all the d_i are different from zero, the linear mapping with matrix $[D]$ is non-singular.*

Indeed, in this case, a linear combination of the vectors B_i can be expressed as follows:

$$\sum \beta_i\,B_i = \sum \beta_i\,d_i\,A_i\,.$$

Since the coefficients d_i are all non-zero and the vectors A_i are linearly independent, this linear combination of the B_i can yield the zero vector θ only if all the β_i are zero. The vectors B_i are therefore linearly independent.

The matrix $[D]$ of the linear mapping D can then be expressed as follows:

$$D = \begin{bmatrix} d_1 & 0 & \dots & 0 \\ 0 & d_2 & \dots & 0 \\ 0 & 0 & \dots & d_n \end{bmatrix}$$

(III-18) *A matrix in which all elements except those on the main diagonal are zero is called a diagonal matrix.*

Diagonal matrices play an important role in the theory of linear systems and in that of quadratic forms.

(III-19) *A diagonal matrix in which all the elements d_i are equal is said to be a scalar matrix.*

Denoting by $[S]$ the scalar matrix in which all diagonal elements are equal to s, we have, for any $X \in E_n$:

$$[S]\, X = sX\,.$$

Scalar matrices have an interesting property. We have seen earlier that a product of matrices is not commutative in general, $[A][B] \neq [B][A]$. However

(III-20) *Scalar $n \times n$ matrices commute with any other $n \times n$ matrix, and they alone have this property.*

Consider a scalar matrix $[S]$ and an arbitrary matrix $[A]$. The general element of the matrix $[S][A]$ is

$$p_i^k = \sum_l s_i^l\, a_l^k = sa_i^k\,, \text{ since } s_i^l = \begin{cases} s & \text{if } i = l\,, \\ 0 & \text{if } i \neq l\,. \end{cases}$$

Similarly, the general element of the matrix $[A][S]$ is

$$q^k = \sum_l a^l\, s_l^k = sa_i^k = p_i^k\,.$$

Hence

$$[S]\,[A] = [A]\,[S]$$

for any $[A]$.

So that scalar matrices commute with any matrix. We now show that they alone have this property. Consider a matrix $[M]$ which commutes with any matrix $[A]$.

For any $[A]$, we have

$$\sum_l a_i^l m_l^k = \sum_l m_i^l a_l^k \qquad \begin{matrix} i = 1, 2, ..., n \\ k = 1, 2, ..., n . \end{matrix}$$

Since this relation must be satisfied for any $[A]$, we choose this matrix such that all its columns are zero except for the kth, which is arbitrary. We can then deduce:

$$a_i^k m_k^k = \sum_l m_i^l a_l^k \qquad i = 1, 2, ..., n .$$

Since this relation must be satisfied for any a_i^k, the coefficients of a_i^k must be identical in both sides, which implies

$$m_i^l = 0 \qquad \text{for} \qquad l \neq i .$$

$$m_i^i = m_k^k$$

In other words, all the elements of the ith row of $[M]$ are zero, except that lying on the main diagonal, which is equal to m_k^k. Letting i assume all the values 1, 2, ..., n, we see that all the elements of $[M]$ are zero, except those on the main diagonal, which are all equal. The matrix $[M]$ is therefore a scalar matrix.

CHAPTER IV

DETERMINANTS

We have seen the fundamental importance of non-singular linear mappings and of the matrices associated with them. But the criteria we gave for non-singularity are generally not easy to apply.

In this chapter we will introduce a numerical matrix function which will give us a simple criterion for non-singularity.

We associate with the square $n \times n$ matrices a scalar function of their column vectors $C_1, C_2, \ldots, C_n$: the determinant.

The determinant function will be defined by three properties which will be completed later by a fourth existence property.

(IV-1) *The determinant associated with a square $n \times n$ matrix is a scalar function of its column vectors $C_1, C_2, \ldots, C_n$ satisfying the following conditions:*

(IV-2) $D(C_1, C_2, \ldots, C_i, C_{i+1}, \ldots, C_n) = - D(C_1, C_2, \ldots, C_{i+1}, C_i, \ldots, C_n)$

(IV-3) $D(C_1, C_2, \ldots, C_i' + C'', \ldots, C_n) = D(C_1, C_2, \ldots, C_i', \ldots, C_n) + $
$$+ D(C_1, C_2, \ldots, C_i'', \ldots, C_n)$$

(IV-4) $D(\alpha_1 C_1, \alpha_2 C_2, \ldots, \alpha_i C_i, \ldots, \alpha_n C_n) =$
$$= \alpha_1 \alpha_2, \ldots, \alpha_i, \ldots, \alpha_n D(C_1, C_2, \ldots, C_i, \ldots, C_n) .$$

Some interesting properties can be deduced from

these axioms.

(IV-5) *If two vectors in the determinant* $D(C_1, C_2, \ldots, C_n)$ *are identical, this determinant is equal to zero.*

We prove this in the case where the two identical vectors are consecutive:

$$C_{i+1} = C_i .$$

The value of the determinant does not change if we interchange C_{i+1} and C_i which are identical, so that applying Axiom (IV-2) yields

$$D = -D ,$$

which implies $D = 0$.

If the identical vectors C_i and C_j are not consecutive, they can be brought together by permuting C_j with C_{j-1} then with C_{j-2}, ..., until C_{i+1}. We have then performed $j - (i + 1)$ permutations and since the determinant D changes sign at each permutation, we have multiplied D by $(-1)^{j-i-1}$. In any case, we get back to the previous case and since the determinant obtained in this way has two identical consecutive vectors C it must be equal to zero.

(IV-6) *COROLLARY. The determinant associated with n linearly dependent vectors* $C_1, C_2, \ldots, C_n$ *is equal to zero.*

Since the vectors C_i are linearly dependent, they satisfy the following relation:

$$\alpha_1 C_1 + \alpha_2 C_2 + \cdots + \alpha_n C_n = \theta ,$$

where the α_i are not all identically zero. Suppose for instance that α_1 is non-zero; we can deduce the expression for C_1 in terms of $C_2, \ldots, C_n$ from the above relation:

$$C_1 = \beta_2 C_2 + \beta_3 C_3 + \cdots + \beta_n C_n ,$$

where

$$\beta_i = - \alpha_i/\alpha_1 .$$

The determinant of the vectors C_i can therefore be expressed as follows:

$$D(C_1, C_2, \ldots, C_n) = \sum_i D(\beta_i C_i, C_2, C_3, \ldots, C_n) = \sum_i \beta_i D(C_i, C_1, C_2, \ldots, C_n)$$

and every determinant in the sum has two identical vectors, the vector C_i and one of the other vectors $C_2, C_3, \ldots, C_n$. These determinants vanish and we have

$$D(C_1, C_2, \ldots, C_n) = 0 .$$

We have just found some properties of determinants, but we have not yet shown that they exist, nor indicated how to construct them. We therefore add the following fourth property to the properties (IV-2) to (IV-4).

(IV-7) *The determinant associated with the unit matrix* [*I*] *is equal to* 1.

In other words:

$$D(e_1, e_2, \ldots, e_i, \ldots, e_n) = 1$$

where e_i is the *i*th column vector of the matrix [*I*], that is, a vector of which all components are zero except for the *i*th one, which is equal to 1.

Note here that the determinant is associated with a matrix and not with a linear mapping. We shall see later that the value of the determinant is uniquely determined by the elements of the matrix.

The column vectors e_i of the matrix $[I]$ form a basis of E_n and therefore the column vectors $C_1, C_2, \ldots, C_n$ of an arbitrary $n \times n$ matrix can be expressed as follows:

$$\begin{aligned} C_1 &= a_1^1\, e_1 + a_2^1\, e_2 + \cdots + a_n^1\, e_n \\ C_2 &= a_1^2\, e_1 + a_2^2\, e_2 + \cdots + a_n^2\, e_n \\ &\vdots \\ C_n &= a_1^n\, e_1 + a_2^n\, e_2 + \cdots + a_n^n\, e_n\,. \end{aligned}$$

If we substitute these values of the C_i into the determinant $D(C_1, C_2, \ldots, C_n)$ and if we use the properties (IV-2) and (IV-3), we decompose D into a sum of n^n determinants, each containing only the vectors e_i. Indeed, if in the determinant D we start by decomposing the vector C_1, we transform D into a sum of n determinants D_i of the form $a_i\, D\,(e_i, C_2, \ldots, C_n)$. And then, decomposing C_2, we transform each determinant $a_i\, D(e_i, C_2, \ldots, C_n)$ into a sum of n determinants of form:

$$a_i^1\, a_j^2\, D(e_i, e_j, C_3, \ldots, C_n)$$

and so on, until the decomposition of C_n. At each stage we obtain n new determinants from each determinant of the previous stage.

At stage n we obtain n^n determinants of form:

$$a_i^1\, a_j^2 \ldots D(e_i, e_j, \ldots)$$

of which only $n!$ are different from zero.

Indeed, we have seen that the decomposition of C_1 yields n determinants

$$a_i^1\, D(e_i, C_2, \ldots, C_n)\,,$$

and the decomposition of C_2 yields n new determinants of the following form from each determinant of the first stage:

$$a_i^1 \, a_j^2 \, D(e_i, e_j, C_3, \ldots, C_n) \, .$$

But among these n determinants, that in which e_j is equal to e_i is zero, since it has two identical vectors; hence there are only $n - 1$ non-zero determinants at stage 2; at stage 3, we see in the same way that there are only $n - 2$ determinants

$$a_i^1 \, a_j^2 \, a_k^3 \, D(e_i, e_j, e_k, \ldots, C_n)$$

which are non-zero, those for which e_k is equal neither to e_i nor to e_j , and so on. We finally get

$$n(n-1)(n-2) \ldots 1 = n\,!$$

non-zero determinants of form

$$a_{r_1}^1 \, a_{r_2}^2 \, a_{r_3}^3 \ldots a_{r_n}^n \, D(e_{r_1}, e_{r_2}, e_{r_3}, \ldots, e_{r_n}) \, ,$$

where the sequence $r_1, r_2, r_3, \ldots, r_n$ denotes a permutation of the indices $1,2,\ldots,n$. The upper indices follow the natural order because of the very way in which the decomposition has been performed.

The lower indices are all different, since if any two of them were identical, the two associated vectors e_{r_i} and e_{r_j} would be identical and the corresponding determinant would be zero.

On the other hand, there are no two identical determinants $D(e_{r_1}, \ldots, e_{r_n})$. Indeed, at each stage i we found the vector e_{r_i} in position i once and only once. It is therefore not possible to find the same sequence $r_1, r_2, \ldots, r_n$ twice. Consequently, since the determinants $D(e_{r_1}, e_{r_2}, \ldots, e_{r_n})$ are all different and since there are $n!$ of them, the indices $r_1, r_2, \ldots, r_n$ cover all the possible permutations of the integers $1,2,\ldots,n$.

It is easy to evaluate the determinants $D(e_{r_1}, ..., e_{r_n})$. Let $k(r_1, r_2, ..., r_n)$ be the number of transpositions necessary to put the indices $r_1, r_2, ..., r_n$ in the natural order $1, 2, ..., n$; using Axiom (IV-7), we get:

$$D(e_{r_1}, e_{r_2}, ..., e_{r_n}) = (-1)^{k(r_1, r_2, ..., r_n)} D(e_1, e_2, ..., e_n)$$
$$= (-1)^{k(r_1, r_2, ..., r_n)}$$

Consequently, the determinant $D(C_1, C_2, ..., C_n)$ can be expressed as follows:

(IV-8) $$D(C_1, C_2, ..., C_n) = \sum_{r_i \in P_n} (-1)^{k(r_1, r_2, ..., r_n)} a^1_{r_1} a^2_{r_2} \dots a^n_{r_n}.$$

Whence the rule:

(IV-8) *A determinant with n rows and n columns is the sum of all the products obtained by taking one term and one only in each row and in each column, each product bearing the sign + if it is necessary to perform an even number of transpositions to put the lower indices in the natural order, the sign - if this number of transpositions is odd.*

In fact, a determinant is rarely computed from this formula which is far too complicated for determinants involving many terms. One rather makes use of the following property:

(IV-9) *A determinant $D(C_1, C_2, ..., C_n)$ is a linear combination of the elements of any one of its columns.*

Indeed, if we express the vector C_j in terms of the basis vectors $e_i, e_2, ..., e_n$ and if we decompose the determinant $D(C_1, C_2, ..., C_n)$ into a sum, we get:

$$D(C_1, C_2, ..., C_j, ..., C_n) = \sum_i a^j_i D(C_1, C_2, ..., e_i, ..., C_n).$$

Computation of the coefficients of the a_i^j, will give us a new evaluation of the determinant $D(C_1, C_2, ..., C_n)$.

We first get

$$D(C_1, C_2, ..., e_i, ..., C_n) = (-1)^{j-1} D(e_i, C_1, C_2, ..., C_{j-1}, C_{j+i}, ..., C_n) .$$

If we decompose the latter determinant as we have done previously in terms of the components of the vectors C, we get $(n - 1)!$ non-zero determinants, since we started expanding it in terms of its second vector, and we then have:

$$(-1)^{j-1} D(e_i, C_1, C_2, ..., C_n) = (-1)^{j-1} \sum (-1)^{k(r_1, r_2, ..., r_n)}$$
$$a_{r_1}^1 a_{r_2}^2 ... a_{r_n}^n D(e_i, e_1, ..., e_{i-1}, e_{i+1}, e_n) .$$

There are no elements of the jth column of the initial determinant in the terms above, since the column C_j has been replaced by a vector e_i and there are no elements of the ith row since the terms of this row would correspond to zero determinants (since they contain two identical vectors). If we now perform $i - 1$ transpositions on the vector e_i we put it back in its normal position and finally get

$$(-1)^{j+1-2} D(C_1, C_2, ..., C_i, ..., C_n) = (-1)^{i+j} \sum (-1)^{k(r_1, r_2, ..., r_n)} a_{r_1}^1 a_{r_2}^2 ... a_{r_n}^n .$$

The sum in the right-hand side is simply a determinant with $n - 1$ rows and $n - 1$ columns obtained from the initial determinant by deleting the ith row and the ith column. And we finally have, denoting this determinant by D_i^j :

(IV-10) $$D(C_1, C_2, ..., C_n) = \sum_i (-1)^{i+j} a_i^j D_i^j .$$

In other words, a determinant of order n is a linear combination of determinants of order $n - 1$. The formula above is said to give the expansion of $D(C_1, C_2, ..., C_n)$ in terms of the elements of the jth column.

The following very useful relation can be deduced from the above formula:

(IV-11)
$$\boxed{\sum_i (-1)^{i+j} a_i^k D_i^j = 0}$$

for $k = 1,2,\ldots,j-1,j+1,\ldots,n$.

We see that this formula is analogous to the formula for the determinant D, except that the vector C_j has been replaced by the vector C_k. Consequently, since the determinant obtained in this way has two identical columns, it is equal to zero.

We have proved earlier that the determinant of n linearly dependent vectors is always zero. *We now show that the determinant of n linearly independent vectors is always non-zero.*

Let $A_1, A_2, \ldots, A_n$ be linearly independent vectors; they form a basis of the space E_n and all the vectors of E_n can be expressed in terms of the A_i, in particular the vectors e_i :

$$\begin{aligned}
e_1 &= \alpha_1^1 A_1 + \alpha_2^1 A_2 + \cdots + \alpha_n^1 A_n\,,\\
e_2 &= \alpha_1^2 A_1 + \alpha_2^2 A_2 + \cdots + \alpha_n^2 A_n\,,\\
&\vdots\\
e_n &= \alpha_1^n A_1 + \alpha_2^n A_2 + \cdots + \alpha_n^n A_n\,.
\end{aligned}$$

In the determinant $D(e_1, e_2, \ldots, e_n)$, the vectors e_i can be replaced by their expression in terms of the A_j. By reasoning analogous to that used to express the value of a determinant $D(C_1, C_2, \ldots, C_n)$ in terms of the components a_i^j, we can express $D(e_1, e_2, \ldots, e_n)$ as follows:

$$D(e_1, e_2, \ldots, e_n) = \sum_{r_1 r_2 \ldots r_n} \alpha_{r_1}^1 \alpha_{r_2}^2 \ldots \alpha_{r_n}^n \, D(A_{r_1}, A_{r_2}, \ldots, A_{r_n}).$$

A suitable number of transpositions $k(r_1, r_2, ..., r_n)$ allows us to set the vectors A_i in the natural order of their indices, which finally yields:

$$D(e_1, e_2, ..., e_n) = \sum_{r_1 r_2 \ldots r_n} (-1)^{k(r_1 r_2 \ldots r_n)} \alpha^1_{r_1} \alpha^2_{r_2} \ldots \alpha^n_{r_n} D(A_1, A_2, ..., A_n)$$

$$\text{(IV-12)} \qquad = D(A_1, A_2, ..., A_n) \sum_{r_1 r_2 \ldots r_n} (-1)^{k(r_1 r_2 \ldots r_n)} \alpha^1_{r_1} \alpha^2_{r_2} \ldots \alpha^n_{r_n} .$$

Now, from the hypotheses concerning determinants, $D(e_1, e_2, ..., e_n)$ is equal to 1, which would be impossible if $D(A_1, A_2, ..., A_n)$ were zero. Hence $D(A_1, A_2, ..., A_n)$ is not equal to zero.

We can therefore state the following theorem, which shows the practical value of the concept of determinant as a criterion of independence for vector systems.

(IV-13) *n vectors of E_n are linearly independent if and only if the associated determinant is not zero.*

We have just seen that the condition is necessary.

We now show that it is sufficient. The determinant D is different from zero. This means that it is impossible to find n scalars which are not identically zero such that

$$\gamma_1 C_1 + \gamma_2 C_2 + \cdots + \gamma_n C_n = \theta$$

since if we could the result (IV-6) would imply that the determinant is zero, which contradicts our hypothesis. The vectors C_i are therefore linearly independent.

In view of its importance we now give two other very useful formulations of this theorem.

(IV-14) *A determinant is different from zero if and*

only if its column vectors are linearly independent.

(IV-15) *A linear mapping of* e_n *into itself is non-singular if and only if the determinant associated with its matrix is non-zero.*

This is true since a mapping is non-singular if and only if the column vectors of its matrix are linearly independent.

We have seen earlier that in order to define the determinant function completely, we set $D(I) = 1$. We then obtained the value of the determinant associated with an arbitrary matrix $D(A)$ assuming that the column vectors $C_1, C_2, \ldots, C_n$ of A were represented in terms of the basis vectors $e_1, e_2, \ldots, e_n$, the column vectors of the matrix I.

We shall see later that the determinant associated with an $n \times n$ matrix is invariant under any change of basis.

We note moreover that since a matrix is defined by its column vectors, we can equally well speak of the determinant $D(A)$ associated with a matrix $[A]$, or of the determinant $D(C_1, C_2, \ldots, C_n)$ associated with the column vectors of the matrix $[A]$. The determinant associated with the matrix $[A]$ is often denoted $|A|$.

DETERMINANTS AS FUNCTIONS OF THE ROW VECTORS

We have defined the determinant associated with a matrix $[a_i^k]$ as being a function of the column vectors $C_1, C_2, \ldots, C_n$. But we could also have considered it as being a function of the row vectors $L_1, L_2, \ldots, L_n$, where the vector L_i has coordinates $a_i^1, a_i^2, \ldots, a_i^n$.

We could have required that the determinant defined

in terms of the row vectors L_i satisfy the same conditions as those we have set above, namely:

$$\Delta = (L_1, L_2, ..., L_{i+1}, L_i, ..., L_n) = -\Delta(L_1, L_2, ..., L_i, L_{i+1}, ..., L_n)$$

$$\Delta = (L_1, L_2, ..., L'_i + L''_i, ..., L_n) = \Delta(L_1, L_2, ..., L'_i, ..., L_n) + \Delta(L_1, L_2, ..., L''_i, ..., L_n)$$

$$\Delta(\alpha_1 L_1, \alpha_2 L_2, ..., \alpha_n L_n) = \alpha_1 \alpha_2 ... \alpha_n \Delta(L_1, L_2, ..., L_n)$$

$$\Delta(e_1, e_2, ..., e_n) = 1 .$$

These conditions allow us to calculate the explicit value of the function Δ. With exactly the same reasoning as before, we get:

$$(L_1, L_2, ..., L_n) = \sum_{r'_1 r'_2 ... r'_n} (-1)^{k(r'_1 r'_2 ... r'_n)} a_1^{r'_1} a_2^{r'_2} ... a_n^{r'_n}$$

There are $n!$ terms of this type, each composed of n factors, with only one factor taken from each row and each column. All these terms are different and consequently we find all the permutations of the upper indices. We therefore find the same terms as in the previous case, since it would be sufficient to rearrange the upper indices in the natural order to find all the permutations of the integers $1,2,...,n$ in the lower indices. We can show moreover that the sign before any given term is always the same in both formulas. It follows that the expressions are identical.

We must therefore show that

$$(-1)^{k(r_1, r_2, ..., r_n)} = (-1)^{k(r'_1, r'_2, ..., r'_n)}$$

for the terms in which we find the same elements, though in a different order.

For these terms:

$$a_{r_1}^1 a_{r_2}^2 ... a_{r_n}^n = a_1^{r'_1} a_2^{r'_2} ... a_n^{r'_n} ,$$

that is, identifying identical terms,

$$a_{r_i}^i = a_j^{r_j},$$

which implies

$$r_i = j, \qquad r_j = i.$$

Therefore, to the term of rank i and value j in the first permutation, there corresponds the term of rank j and value i in the second. *The two permutations are said to be conjugate.*

Now the number $k(r_1, r_2, \ldots, r_n)$ denotes the number of permutations it is necessary to perform on the n integers $r_1, r_2, \ldots, r_n$ between 1 and n in order to rearrange them in the natural order. We now show that this number is the same for two conjugate permutations.

To prove, this, we introduce the concept of inversion. Two terms in a sequence $r_1, r_2, \ldots, r_n$ are said to form an "inversion" if they are not in the natural order; the number of inversions of a term is thus the number of terms which follow it but would precede it in the natural order. The number of inversions of a sequence is the sum of the inversions of its terms. Under these conditions:

(IV-16) *The number of transpositions necessary to rearrange a permutation $r_1, r_2, \ldots, r_n$ in the natural order is equal to its number of inversions.*

We first prove a lemma that will be used in the course of the proof:

(IV-17) *In a sequence $r_1, r_2, \ldots, r_n$, the concept of inversion is transitive. In other words, if there is no inversion between r_i and r_j nor between r_j and $r_k (i < j < k)$, then there is no inversion between r_i and r_k.*

There is no inversion between two elements r_α and r_β if and only if

$$r_\alpha < r_\beta, \quad \text{where} \quad \alpha < \beta.$$

Therefore, if there is no inversion between r_i and r_j, or between r_j and r_k, we must have

$$r_i < r_j \quad \text{and} \quad r_j < r_k,$$

whence

$$r_i < r_k.$$

There is therefore no inversion between r_i and r_k.

We now consider the sequence $r_1, r_2, ..., r_n$ and then the pairs $r_1 r_2, r_3 r_4, ..., r_{n-1} r_n$ or $r_{n-2} r_{n-1}, r_n$ according as n is even or odd.

We perform a transposition on every pair presenting an inversion; if there are p_1 such pairs, we have removed p_1 inversions in p_1 transpositions without introducing any new inversions, and we get a new permutation

$$r'_1 r'_2 ... r_n.$$

If this permutation reproduces the natural order, the theorem is proved, since the sequence had p inversions at the beginning and these have been removed in p transpositions.

If the permutation is different from the natural order, we form the pairs

$$r'_1, \quad r'_2 r'_3, \quad r'_4 r'_5, ... r_{n-1} r_n$$

$$r_{n-2} r_{n-1}, \quad r_n$$

according as n is even or odd.

Some of these pairs must contain inversions. Otherwise, the fact that there is no inversion between r_1' and r_2' and that there would be none between r_2' and r_3' would imply that there would be none between r_1' and r_3' and so forth. We would find no inversions in the sequence, which contradicts the hypothesis that the sequence r_i' is not in the natural order.

Hence, if there are p_2 pairs that form inversions, we can remove these p_2 inversions by means of p_2 transpositions without creating any new inversions and we have performed altogther $p_1 + p_2$ transpositions, removing $p_1 + p_2$ inversions.

We continue this process until there are no inversions left in the last decomposition into pairs. The sequence is then in natural order and we have removed its $p_1 + p_2 + \cdots + p_k$ inversions in $p_1 + p_2 + \cdots + p_k$ transpositions.

We are now going to prove the result stated earlier:

(IV-18) *Two conjugate permutations have the same number of inversions.*

Let us consider two conjugate permutations

$$r_1\, r_2 \ldots r_i\, r_n$$

$$r_1'\, r_2' \ldots r_j\, r_n'$$

If k is in position i in the first one, i will be in position k in the second, etc.; for instance, the two following permutations are conjugate, as can readily be seen:

$$\begin{matrix} 3 & 6 & 2 & 9 & 4 & 5 & 1 & 7 & 8 \\ 7 & 3 & 1 & 5 & 6 & 2 & 8 & 9 & 4 \end{matrix}$$

Let us then consider two terms forming an inversion in the first sequence, say r_i and r_j ; this means that $i < j$ and $r_i > r_j$.

In the second sequence, i will be in position r_i and j in position r_j ; since r_i is greater than r_j, these two terms will be in the order j, i, and since j is greater than i, they will also form an inversion. Hence, if two terms form an inversion in the first sequence, their homologues will form an inversion in the second. And we can show similarly that if two terms do not form an inversion in the first sequence, they do not form one in the second.

Consequently, the number of inversions is the same in both sequences.

We have thus proved that

$$k(r_1, r_2, ..., r_n) = k(r'_1, r'_2. ..., r'_n)$$

for two terms containing the same factors in the expressions for $D(C_1, C_2, ..., C_n)$ and for $\Delta(L_1, L_2, ..., L_n)$, and therefore:

$$\Delta(L_1, L_2, ..., L_n) = D(C_1, C_2, ..., C_n) .$$

We therefore no longer use the symbol Δ and write

$$D(L_1, L_2, ..., L_n) = D(C_1, ..., C_2, C_n) .$$

We now can state the following result:

(IV-19) *The determinant associated with a square matrix can be considered both in terms of the column vectors* $C_1, C_2, ..., C_n$ *and in terms of the row vectors* $L_1, L_2, ..., L_n$.

Just as we have been able to expand the determinant in terms of the elements of a column, it is also possible to expand it in terms of the elements of a row, which gives the following relation:

$$D(L_1, L_2, \ldots, L_n) = \sum_j (-1)^{i+j} a_i^j D_i^j$$

and implies the following important relation

$$\text{(IV-20)} \qquad \sum_j (-1)^{i+j} a_k^j D_i^j = 0 \qquad \text{for} \qquad k \neq i\,.$$

SOME REMARKS CONCERNING DETERMINANTS

One of the most important uses of determinants is to translate in an operational and algebraic way a concept that was hitherto purely vectorial and abstract, that of the independence of systems of n vectors in E_n and of the non-singularity of linear mappings; the non-singularity of a linear mapping proceeds from the non-vanishing of an algebraic expression which is completely defined in terms of the elements of the associated matrix. But two important restrictions reduce the usefulness of determinants.

They display the independence of systems of n vectors without giving any information about the systems of p vectors, where $p < n$. On the other hand, the results above have been obtained for a special basis of E_n, the vectors $e_1, e_2, \ldots, e_n$. In fact, these two restrictions will vanish thanks to the results we will now establish, and we shall see later that determinants allow us to characterize the rank of a linear mapping or of an arbitrary system of vectors, and this for any basis in E_n.

We now establish, for systems of r vectors of E_n and for $m \times n$ matrices of rank r, theorems analogous to those just established for systems of n vectors of E_n and for non-singular matrices of E_n .

First of all, a definition:

A submatrix of a matrix [A] is any matrix obtained by deleting certain rows and columns from [A].

We then have the following theorem:

(IV-21) *The $m \times n$ matrix [A] has rank r if and only if it is possible to extract from it a square submatrix with r rows and r columns at most whose determinant is not equal to zero.*

The condition is necessary.

The matrix $[A]$ is of rank r. This means that it has r linearly independent column vectors, say

$$f(A_1), f(A_2), \ldots, f(A_r).$$

We now remove from $[A]$ the $n - r$ column vectors $f(A_{r+1}), f(A_{r+2}), \ldots, f(A_n)$. We get a rectangular submatrix with m rows and r columns.

It has r linearly independent column vectors: by Theorem (II-14) it must also have r linearly independent row vectors; in this rectangular submatrix we delete the $m - r$ row vectors which are not linearly independent.

We get a square submatrix with r rows and r columns which has r linearly independent rows; its r columns are therefore linearly independent, so that its determinant is non-zero. And we could not have found a square matrix with more than r rows and r columns satisfying this property.

The condition is sufficient.

We have been able to extract from the matrix $[A]$ a submatrix with r rows and r columns whose determinant is non-zero. This means that its r row vectors and its r column vectors are linearly independent.

We now complete this square matrix with $m - r$ rows by the r missing elements to reconstitute a submatrix with r column vectors of the matrix $[A]$.

The r column vectors of this submatrix are linearly independent. We now complete it with the $n - r$ missing column vectors to reconstitute the matrix $[A]$. There are no more than r linearly independent vectors among the n reconstituted column vectors A_i, since if there were, reversing the operation would yield a square submatrix of $[A]$ with more than r rows and r columns whose determinant would not be equal to zero, which contradicts our hypothesis that r is the greatest number such that a square submatrix of $[A]$ with r rows and r columns has a non-zero determinant.

This result is true in particular for square matrices.

The following result deals with systems of vectors of E_n:

(IV-22) *A system of p vectors* $V_1, V_2, \ldots, V_p \, (p \leqslant n)$ *of* E_n *has rank r* $(r \leqslant p)$ *if and only if it is possible to extract from the matrix* $[V]$ *whose column vectors are the vectors* V_i *a matrix with at most r rows and r columns whose determinant is non-zero.*

The condition is necessary.

Consider the system of p vectors $V_1, V_2, \ldots, V_p$ of rank r. This means that there are r linearly independent vectors among the p vectors under consideration. Suppose, without loss of generality, and to simplify the notation, that these are the vectors $V_1, V_2, \ldots, V_r$. The corresponding matrix $[V_r]$

$$[V_r] = \begin{bmatrix} v_1^1 & v_1^2 & \dots & v_1^r \\ v_2^1 & v_2^2 & \dots & v_2^r \\ \vdots & & & \\ v_n^1 & v_n^2 & \dots & v_n^r \end{bmatrix}$$

has r linearly independent columns and therefore also linearly independent row vectors. We now remove from $[V_r]$ the $n - r$ rows which are not linearly independent. This yields a matrix with r rows and r columns whose r rows are linearly independent, so that its determinant is not zero. We have thus derived from the initial matrix a matrix with r rows and r columns whose determinant is non-zero. It would not have been possible to extract a matrix with more than r rows and r columns having a non-zero determinant. Indeed, if we had introduced one or more additional columns into the matrix $[V_r]$, this matrix would, in any case, still contain only r linearly independent columns, and hence only r linearly independent rows. It follows that the matrix with more than r rows and r columns that could have been derived would have its determinant equal to zero.

The condition is sufficient.

We have been able to extract form the matrix $[V]$ a matrix $[V_r]$ with r rows and r columns whose determinant is non-zero and we could not have extracted a matrix with more than r rows and columns whose determinant is non-zero. We complete this matrix $[V_r]$ with $n - r$ rows whose elements reconstitute the corresponding r vectors V_i. We get a new matrix $[V_r^n]$ which still has only r linearly independent rows. Suppose indeed that it had $r'(r' > r)$ such rows. It would follow that by completing the matrix $[V_r^n]$ with $r' - r$ column vectors, from the V_i which do not already occur in V_r , we would obtain a matrix with r' rows and r' columns for which the r' rows are linearly independent. The determinant of such a matrix would be different from zero, which would contradict our hypothesis that it is not possible to extract from the matrix $[V]$ a matrix with more than r rows and columns whose determinant is non-zero.

The matrix $[V_r^n]$ has r linearly independent rows and consequently r linearly independent columns. If we complete the matrix $[V_r^n]$ with the $p - r$ other vectors of the system under consideration, the new matrix obtained in this way still has r linearly independent columns and cannot have $r''(r'' \geq r)$ of them, for the same reason as above. (Otherwise we could have extracted from the matrix $[V]$ a matrix with r'' rows and columns whose determinant is non-zero, which contradicts our hypothesis.) It follows that there are r linearly independent vectors among the p vectors of the system and the system is indeed of rank r.

We have thus completely defined a scalar criterion allowing us to evaluate the rank of systems of vectors and that of matrices in an n-dimensional space.

The results we have just obtained allow us to clarify certain points not examined in detail earlier.

We have proved that a determinant can be considered both in terms of the row vectors and in terms of the column vectors of the associated matrix. This can be expressed as follows:

(IV-23) *The determinant associated with the transpose of a matrix $[A]$ is equal to the determinant associated with this matrix:*

$$|A'| = |A|.$$

(IV-24) *The determinant of the product of two $n \times n$ matrices is equal to the product of their determinants:*

$$|AB| = |A||B|.$$

Suppose that the two determinants are non-zero. If one of them were zero, $|A|$ for instance, the product $|A|\ |B|$ would be zero. But the fact that $|A|$ is zero means that the matrix $[A]$ is singular, and therefore that the images Y of the transforms of vectors X of E_n form a subspace E_m of E_n (where $m < n$). It follows that the images of the Y under the transformation with matrix $[B]$ all belong to E_m or to a subspace of E_m. In any case, these images belong to a subspace of E_n and the matrix $[B][A]$ is therefore singular, its determinant is zero and the given equality is valid.

We now consider the case where the matrices $[A]$ and $[B]$ are non-singular.

From the vectorial definition of the determinant, we can say:

$$|B| = D(B_1, B_2, \ldots, B_n) = D[g(e_1), g(e_2), \ldots, g(e_n)]$$

and also

$$|A\ B| = D[fg(e_1), fg(e_2), \ldots, fg(e_n)].$$

Since the mapping g is non-singular, the vectors $g(e_1), g(e_2), \ldots, g(e_n)$ form a basis of E_n, and the vectors $fg(e_1), fg(e_2), \ldots, fg(e_n)$ can therefore be expressed in terms of $g(e_1), g(e_2), \ldots, g(e_n)$:

$$fg(e_i) = \lambda_1^i\, g(e_1) + \lambda_2^i\, g(e_2) + \cdots + \lambda_n^i\, g(e_n)\,.$$

We now substitute these values of the $fg(e_i)$ in the determinant $|AB|$, which we then expand following the method used on page 74. After performing the reductions, we see that the determinant can be expressed as follows:

$$D[fg(e_1), fg(e_2), \ldots, fg(e_n)] =$$

$$D[g(e_1), g(e_2), \ldots, g(e_n)] \sum (-1)^{k(r_1, r_2, \ldots, r_n)}\, \lambda_{r_1}^1\, \lambda_{r_2}^2 \ldots \lambda_{r_n}^n\,,$$

that is: $|AB| = \mu_1 |B|$.

Similarly, it is possible to express $fg(e_1), fg(e_2), \ldots, fg(e_n)$ in terms of the vectors $f(e_1), f(e_2), \ldots, f(e_n)$, which form a basis of E_n, and by identical reasoning we get:

$$|AB| = \mu_2 |A|.$$

This means that $|AB|$ is of the following form:

$$|AB| = \mu |A| |B|$$

and we must determine the proportionality factor. Now the terms of $|AB|$ are linear with respect to each of the terms a_i^j and b_l^m, so that μ depends neither on A nor on B and can therefore be calculated for any value of the matrices $[A]$ or $[B]$. Let us then take $[A] = [I]$; we get

$$|IB| = |B| = \mu |I| |B|$$

Therefore $\mu = 1$ and

$$|AB| = |A| |B|.$$

We now go back to the definition of inverse matrices as begun on page 54. We saw that to any non-singular matrix $[A]$ there corresponds a non-singular matrix $[B]$ such that

$$[B][A] = [I]$$

and we denoted this matrix $[B]$ by $[A]^{-1}$.

If we denote by b_i^k the general element of the matrix $[A]^{-1}$, the relation given above allows us to determine the b_i^k by the following conditions:

$$\sum_j b_i^j a_j^k = \begin{cases} 1 \text{ for } i = k, \\ 0 \text{ for } i \neq k. \end{cases}$$

We see that if we take

$$b_i^j = A_j^i / | A |$$

the above system of n^2 equations in n^2 unknowns b_i^j is satisfied. Indeed, each relation where $i = k$ then represents the expansion of the determinant of the matrix $[A]$ with respect to the elements of its kth column and each relation where $i \neq k$ represents the expansion of an analogous determinant with two identical columns, that is, with zero determinant.

The solution is unique. Indeed, if there were another solution, another matrix would correspond to it, which is impossible since a unique matrix corresponds to a non-singular linear mapping when the basis of the space remains the same. The matrix $[A]^{-1}$ is unique, so that the coefficients b_i^k are also unique.

(IV-25) *The matrix* $[A]^{-1}$ *is non-singular.*

We have indeed

$$| [A] [A]^{-1} | = |I| = 1 .$$

Hence

$$| [A]^{-1} | \neq 0 .$$

Moreover

$$| [A]^{-1} | = \frac{1}{| A |} .$$

Theorem (IV-24) has an interesting corollary.

(IV-26) *The determinant* M *associated with the matrix* $[M]_{n \times n}$ *of a linear mapping of* E_n *into* E_n *is invariant under any change of basis.*

We have indeed, from relation (III-11),

$$[M_B] = [B]^{-1}[M_A][B],$$

where $[B]$ is the matrix of the change of basis. Considering the determinants and applying Theorem (IV-24), we find:

$$|M_B| = \frac{1}{|B|}|M_A||B| = |M_A|.$$

We show in this way that the results obtained previously are independent of the basis in the space under consideration.

CHAPTER V

SYSTEMS OF LINEAR EQUATIONS

One of the most useful applications of the methods we have examined in the first three chapters is the solution of systems of linear equations, which we will now consider as an illustration of vector methods.

Consider the most general system of linear equations, namely:

$$\begin{aligned} a_1^1 x_1 + a_1^2 x_2 + \cdots + a_1^n x_n &= b_1 \\ a_2^1 x_1 + a_2^2 x_2 + \cdots + a_2^n x_n &= b_2 \qquad m \neq n \\ &\vdots \\ a_m^1 x_1 + a_m^2 x_2 + \cdots + a_m^n x_n &= b_m . \end{aligned}$$

This numerical system is equivalent to the following vector system:

$$[A] X = B$$

where $[A]$ is the matrix

$$[A] = \begin{bmatrix} a_1^1 & a_1^2 & \ldots & a_1^n \\ \vdots & & & \\ a_m^1 & a_m^2 & \ldots & a_m^n \end{bmatrix}$$

and X and B are column vectors:

$$X = \begin{bmatrix} x_1 \\ x_2 \\ \vdots \\ x_n \end{bmatrix}, \qquad B = \begin{bmatrix} b_1 \\ b_2 \\ \vdots \\ b_m \end{bmatrix} \qquad X \in E_n, \qquad B \in E_m .$$

We have *solved* such a system if we have found a set of numbers $x_1, x_2, \ldots, x_n$ which, when substituted in the various equations, make the two sides equal.

If the right-hand sides $b_1, b_2, \ldots, b_m$ are all zero, the system is said to be homogeneous.

Before solving this most general system, we are going to solve simpler systems which will moreover give us an idea of the methods to use.

CRAMER SYSTEMS

Consider the following system:

$$\text{(V-1)} \qquad \begin{cases} a_1^1 x_1 + a_1^2 x_2 + \cdots + a_1^n x_n = b_1 , \\ a_2^1 x_1 + a_2^2 x_2 + \cdots + a_2^n x_n = b_2 , \\ \vdots \\ a_n^1 x_1 + a_n^2 x_n + \cdots + a_n^n x_n = b_n ; \end{cases}$$

with n equations and n unknowns. It is equivalent to the following vector relation:

$$[A]X = B$$

where $[A]$ is the matrix of a linear mapping of E_n into E_n. Suppose moreover that this mapping is non-singular, that is, that its range is all of E_n. The relation $[A]X = B$ is then possible for any vector B since every vector of E_n can be considered as the image of another vector X of E_n. Such a system is called a Cramer system.

Since the mapping with matrix $[A]$ is non-singular, it has an inverse mapping with matrix $[A]^{-1}$, and we immediately obtain the following relation:

$$[A]^{-1}[A]X = [A]^{-1}B,$$

that is,

$$\boxed{X = [A]^{-1}B}\ .$$

It follows that the system of numbers $x_1, \ldots, x_n$ which is a solution of the system (V-1) will consist of the components of the vector $[A]^{-1}B$. Therefore, using the expressions for the elements of the matrix $[A]^{-1}$ in terms of the elements of the matrix $[A]$, we have the following result:

$$x_j = \sum_i \frac{A_i^j b_i}{|A|} = \frac{A_j}{|A|}\ .$$

The sum

$$\sum_i A_i^j b_i$$

is a determinant derived from the determinant of the matrix $[A]$ by replacing the vector C_j by the vector B.

Since the vector X is unique, as we have seen earlier, it follows that its components are also unique. We have thus the following result:

(V-2) *A system of n equations in n unknowns such that the determinant of the coefficients is non-zero admits a unique system of solutions given by the following formulas:*

$$x_j = \sum_i \frac{A_i^j b_i}{A}.$$

SYSTEM OF n EQUATIONS IN n UNKNOWNS IN WHICH THE DETERMINANT OF THE COEFFICIENTS IS ZERO

Consider a system analogous to the system above:

$$[A]X = B.$$

But suppose also that the determinant associated with the matrix $[A]$ is zero:

$$|A| = 0.$$

This means that the linear mapping f with matrix $[A]$ is singular and consequently that its range is not all of the space E_n. It follows that the equations (V-1) admit a solution only in the case where the vector B belongs to the range of the mapping f. More precisely:

(V-3) *The equation* $[A]X = B$ *admits a system of solutions if and only if the vector* B *belongs to the range of the linear mapping* f *with matrix* $[A]$.

The condition is necessary. The system admits a solution. This means that there is a vector X_0 such that:

$$[A]X_0 = B.$$

B must therefore belong to the range of the mapping f.

The condition is sufficient. The vector B belongs to the range of the mapping f. It follows that B can be expressed as follows:

$$B = x_1 f(A_1) + x_2 f(A_2) + \cdots + x_n f(A)_n$$

that is, using the properties of linear mappings,

$$B = f(x_1 A_1 + x_2 A_2 + \cdots + x_n A_n) = f(X),$$

where X denotes the following vector of E_n:

$$X = x_1 A_1 + x_2 A_2 + \cdots + x_n A_n .$$

And the system admits a solution, since we have found a vector X satisfying the relation

$$[A] X = B .$$

Suppose that the matrix $[A]$ is of rank r. This means that its range is an r-dimensional subspace of E_n

The condition that B must belong to this subspace can be expressed as follows:

$$B = b_1' f(A_1) + b_2' f(A_2) + \cdots + b_r' f(A_r),$$

where $f(A_1), f(A_r), \ldots, f(A_n)$ are the basis vectors of the range of the mapping f.

The condition is not explicit in this form and we therefore express it in another way:

$$
\begin{aligned}
E_1 &= a_1^1 x_1 + a_1^2 x_2 + \cdots + a_1^n x_n = b_1 , \\
E_2 &= a_2^1 x_1 + a_2^2 x_2 + \cdots + a_2^n x_n = b_2 , \\
&\cdots\cdots\cdots\cdots\cdots\cdots\cdots\cdots \\
E_n &= a_n^1 x_1 + a_n^2 x_2 + \cdots + a_n^n x_n = b_n .
\end{aligned}
$$

Since the matrix $[A]$ of the system is of rank r, it contains r linearly independent row vectors, $L_1, L_2, \ldots, L_r$ say. Each row vector L_{r+1}, L_{r+2}, L_n is therefore a linear combination of $L_1, L_2, \ldots, L_r$; for instance

$$L_s = \lambda_1^s L_1 + \lambda_2^s L_r + \cdots + \lambda_r^s E_r \qquad s > r ,$$

that is, in terms of the components of the vector L_s:

$$
\begin{aligned}
a_s^1 &= \lambda_1^s a_1^1 + \lambda_2^s a_2^1 + \cdots + \lambda_r^s a_r^1 , \\
a_s^2 &= \lambda_1^s a_1^2 + \lambda_2^s a_2^2 + \cdots + \lambda_r^s a_r^2 , \\
&\vdots \\
a_s^n &= \lambda_1^s a_1^n + \lambda_2^s a_2^n + \cdots + \lambda_r^s a_r^n .
\end{aligned}
$$

Substituting these values in the sth equation of the system:

$$E_s = a_s^1 x_1 + a_s^2 x_2 + \cdots + a_s^n x_n = b_s ,$$

we get

$$
\begin{aligned}
E_s &= (\lambda_1^s a_1^1 + \lambda_2^s a_2^1 + \cdots + \lambda_r^s a_r^1)\, x_1 = (\lambda_1^s a_1^2 + \lambda_2^s a_2^2 + \cdots + \lambda_r^s a_r^2)\, x_2 + \cdots \\
&\quad + (\lambda_1^s a_1^n + \lambda_2^s a_2^n + \cdots + \lambda_r^s a_r^n)\, x_n = \lambda_1^s(a_1^1 x_1 + a_1^2 x_2 + \cdots + a_1^n x_n) + \\
&\quad + \lambda_2^s(a_2^1 x_1 + a_2^2 x_2 + \cdots + a_2^n x_n) + \cdots + \lambda_r^s(a_r^1 x_1 + a_r^2 x_2 + \cdots + a_r^n x_n) \\
&= \lambda_1^s E_1 + \lambda_2^s E_2 + \cdots + \lambda_r^s E_r ,
\end{aligned}
$$

or, considering the right-hand sides:

$$b_s = \lambda_1^s b_1 + \lambda_2^s b_2 + \cdots + \lambda_r^s b_r \qquad \text{for } s = r+1, \quad r+2, \ldots, n.$$

We have thus proved the following result:

(V-4) *In a system of rank r with n equations and n unknowns, the n-r last equations of the system are consequences of the r first ones. Therefore:*
The system has a solution if and only if the last n-r right-hand sides satisfy the same relations as the last n-r left-hand sides. If this is the case, it is sufficient to solve the system formed by the r first equations of the initial system, this latter system being then automatically satisfied.

We are led back to the solution of a system of r equations in n unknowns, where the number of unknowns is greater than the number of equations.

We will not solve the system completely now, but consider a more general system of equations whose solution can be reduced to the solution of the system above. At the same time, we shall determine even more accurately the conditions which must be satisfied by the right-hand sides $b_1, b_2, \ldots, b_n$.

Consider the following most general system of m equations in n unknowns, where the matrix of the coefficients is of rank r:

$$\text{(V-5)} \qquad \begin{cases} E_1 = a_1^1 x_1 + a_1^2 x_2 + \cdots + a_1^n x_n = b_1, \\ E_2 = a_2^1 x_1 + a_2^2 x_2 + \cdots + a_2^n x_n = b_2, \\ \vdots \\ E_m = a_m^1 x_1 + a_m^2 x_2 + \cdots + a_m^n x_n = b_m. \end{cases}$$

This system is analogous to the following system:

$$[A]X = B \qquad X \in E_n, \qquad B \in D_r \subset E_m.$$

The main difference from the case we studied earlier is that the linear mapping f with matrix $[A]$ makes vectors of E_m correspond to vectors of E_n.

All the results established above for systems of n equations in n unknowns, where the matrix of the coefficients is singular, remain valid. But we are going to state the conditions for existence of the solutions in a more obvious and manageable form.

We first introduce some definitions:

(V-6) *A principal determinant of the system* (V-5) *considered above is a non-zero determinant with r rows and r columns extracted from the matrix* $[A]$.

By Theorem (IV-21), such a determinant can always be extracted from the matrix $[A]$ since the system is of rank r, and it is not possible to extract a non-zero determinant with more than r rows and r columns.

Suppose, to simplify notations, that the principal determinant of the system is we established on page 99 for

$$|A_r| = \begin{vmatrix} a_1^1 & a_1^2 & \dots & a_1^r \\ a_2^1 & a_2^2 & \dots & a_2^r \\ \vdots & & & \\ a_r^1 & a_r^2 & \dots & a_r^r \end{vmatrix}.$$

(V-7) *A characteristic determinant of the system is any determinant* D_k *deduced from* $|A_r|$ *as follows:* $|A_r|$ *is completed with an* $(r + 1)$*th column consisting of the elements* $b_1, b_2, \ldots, b_k$ *and an* $(r + 1)$*th row* L_k *consisting of the elements* $a_k^1, a_k^2, \ldots, a_k^r, b_k$ *where* $k > r$.

There exist therefore $m - r$ characteristic determinants and we have the following result:

(V-8) *The system* (V-5) *admits a solution if and only if the* $m - r$ *characteristic determinants are zero.*

The condition is necessary. Since the system admits a solution, this means that B belongs to the range D of the mapping with matrix $[A]$. Consider then the following matrix:

$$[\mathcal{A}] = \begin{bmatrix} a_1^1 & a_1^2 & \cdots & a_1^n & b_1 \\ a_2^1 & a_2^2 & \cdots & a_2^n & b_2 \\ \cdot & \cdot & \cdot & \cdot & \cdot \\ a_m^1 & a_m^2 & \cdots & a_m^n & b_m \end{bmatrix}.$$

It is of rank r since B is a linear combination of the r basis vectors of D. Hence, if we delete the column vectors $A_{r+1}, A_{r+2}, \ldots, A_n$, we again obtain a matrix of rank r. If, in this new matrix, we delete the rows $L_{r+1}, L_{r+2}, \ldots, L_m$, keeping L_k, we again obtain a square matrix of rank r with $r + 1$ rows and $r + 1$ columns; its determinant is thus zero for any k.

The condition is sufficient. All the characteristic determinants are equal to zero. This means that we have the following relations:

$$L_k = \lambda_1^k L_1 + \lambda_2^k L_2 + \cdots + \lambda_r^k L_r \qquad k = r + 1, \ldots, m.$$

Indeed, since the principal determinant of the system is not equal to zero, the r first row vectors of the characteristic determinants are linearly independent and the $(r + 1)$th is a linear combination of the r first ones. Consequently, the matrix $[A]$, whose $m - r$ last row vectors are linear combinations of the r first ones, is of rank r, and since its r first column vectors are linearly independent, it follows that its last column vector B is a linear combination of the vectors $A_1, A_2, \ldots, A_r$; hence it belongs to the range of the linear mappings with matrix $[A]$.

Here again we have $m - r$ conditions imposed upon the components of the vector B. They are strictly analogous to those which we established on page 99 for systems of n equations in n unknowns. We can also deduce, as before, that the $m - r$ last equations of the system (V-5) are satisfied if the r first ones are satisfied, provided the system is solvable. We are therefore led to consider--as in the case of systems of n equations in n unknowns where the matrix is of rank r--the following system, which is equivalent to the system (V-5):

$$\text{(V-5a)} \qquad \begin{cases} a_1^1 x_1 + a_1^2 x_2 + \cdots + a_1^n x_n = b_1 , \\ a_2^1 x_1 + a_2^2 x_2 + \cdots + a_2^n x_n = b_2 , \\ \vdots \\ a_r^1 x_1 + a_r^2 x_2 + \cdots + a_r^n x_n = b_r . \end{cases}$$

This is a system of r equations in n unknowns, where $n > r$. This means that we can take arbitrary values for $n - r$ unknowns, and the r other unknowns are then completely defined. To verify this, it is sufficient to express x_1 in terms of $x_2, x_3, \ldots, x_n$ in the first equation. This value of x_1 is then substituted in the $r - 1$ other equations. We get thus a system of

$r - 1$ equations in $n - 1$ unknowns. Repetition of this process yields an equation in $n - r + 1$ unknowns after $r - 1$ operations. We then express one unknown in terms of $n - r$ others, which are thus completely arbitrary.

We therefore take $x_{r+1}, \ldots, x_n$ as parameters and rewrite the system (V-5a) as follows:

$$\begin{aligned}
a_1^1 x_1 + a_1^2 x_2 + \cdots + a_1^r x_r &= b_1 - a_1^{r+1} x_{r+1} - \cdots - a_1^n x_n, \\
a_2^1 x_1 + a_2^2 x_2 + \cdots + a_2^r x_r &= b_2 - a_2^{r+1} x_{r+1} - \cdots - a_2^n x_n, \\
a_r^1 x_1 + a_r^2 x_2 + \cdots + a_r^r x_r &= b_r - a_r^{r+1} x_{r+1} - \cdots - a_r^n x_n.
\end{aligned}$$

This is a Cramer system since its determinant (the principal determinant of the system (V-5)) is not equal to zero. Consequently, it admits the following solution:

$$\text{(V-9)} \qquad \boxed{x_j = \sum_i \frac{A_i^j}{|A_r|}\left(b_i - \sum_{r+1}^{n} a_i^k x_k\right)} \, .$$

Setting

$$\alpha_k^j = \begin{vmatrix} a_1^1 & a_1^2 & \ldots & a_1^j & \ldots & a_1^r \\ a_2^1 & a_2^2 & \ldots & a_2^j & \ldots & a_2^r \\ \vdots & & & & & \\ a_r^1 & a_r^2 & \ldots & a_r^j & \ldots & a_r^r \end{vmatrix} \qquad \begin{array}{l} j \geqslant r+1 \\ 1 \leqslant k \leqslant r \end{array}$$

kth column

we find the following formulas for the general solution of the system:

$$x_1 = y_1 - \frac{1}{|A_r|}(\alpha_1^{r+1} x_{r+1} + \alpha_1^{r+2} x_{r+2} + \cdots + \alpha_1^n x_n),$$

$$x_2 = y_2 - \frac{1}{|A_r|}(\alpha_r^{r+1} r_{r+1} + \alpha_2^{r+2} x_{r+2} + \cdots + \alpha_2^n x_n),$$

$$x_r = y_r - \frac{1}{|A_r|}(\alpha_r^{r+1} x \quad + \alpha_r^{r+2} x \quad + \cdots + \alpha_r^n x_m).$$

EXAMPLE

We wish to solve the system

$$\begin{aligned} 3x_1 + x_2 + 2x_3 + 5x_4 &= 8, \\ 5x_1 + 2x_2 + x_3 + 2x_4 &= 7, \\ x_1 + 4x_2 + 3x_3 + x_4 &= 4, \\ 7x_1 + 3x_2 + x_3 + 5x_4 &= 5. \end{aligned}$$

We have here:

$$|A| = \begin{vmatrix} 3 & 1 & 2 & 5 \\ 5 & 2 & 1 & 2 \\ 1 & 4 & 3 & 1 \\ 7 & 3 & 1 & 5 \end{vmatrix},$$

or, expanding A in terms of the elements of the first column:

$$|A| = 3\begin{vmatrix} 2 & 1 & 2 \\ 4 & 3 & 1 \\ 3 & 1 & 5 \end{vmatrix} - 5\begin{vmatrix} 1 & 2 & 5 \\ 4 & 3 & 1 \\ 3 & 1 & 5 \end{vmatrix} + 1\begin{vmatrix} 1 & 2 & 5 \\ 2 & 1 & 2 \\ 3 & 1 & 5 \end{vmatrix} - 7\begin{vmatrix} 1 & 2 & 5 \\ 2 & 1 & 2 \\ 4 & 3 & 1 \end{vmatrix}$$

$$= 3 \times 1 - 5 \times (-45) + 1 \times (-10) - 7 \times 17$$

$$= 99.$$

We then have:

$$D_1 = \begin{vmatrix} 8 & 1 & 2 & 5 \\ 7 & 2 & 1 & 2 \\ 4 & 4 & 3 & 1 \\ 5 & 3 & 1 & 5 \end{vmatrix},$$

that is,

$$D_1 = 8 \times 1 - 7(-45) + 4 \times (-10) - 5 \times 17 = 198.$$

Similarly

$$D_2 = \begin{vmatrix} 3 & 8 & 2 & 5 \\ 5 & 7 & 1 & 2 \\ 1 & 4 & 3 & 1 \\ 7 & 5 & 1 & 5 \end{vmatrix}$$

or

$$\begin{aligned} D_2 &= 3\begin{vmatrix} 7 & 1 & 2 \\ 4 & 3 & 1 \\ 5 & 1 & 5 \end{vmatrix} - 5\begin{vmatrix} 8 & 2 & 5 \\ 4 & 3 & 1 \\ 5 & 1 & 5 \end{vmatrix} + 1\begin{vmatrix} 8 & 2 & 5 \\ 7 & 1 & 2 \\ 5 & 1 & 5 \end{vmatrix} - 7\begin{vmatrix} 8 & 2 & 5 \\ 7 & 1 & 2 \\ 4 & 3 & 1 \end{vmatrix} \\ &= 3 \times 61 - 5 \times 27 + 1 \times (-26) - 7 \times 47 \\ &= -297. \end{aligned}$$

We also have:

$$D_3 = \begin{vmatrix} 3 & 1 & 8 & 5 \\ 5 & 2 & 7 & 2 \\ 1 & 4 & 4 & 1 \\ 7 & 3 & 5 & 5 \end{vmatrix},$$

or

$$D_3 = 3\begin{vmatrix} 2 & 7 & 2 \\ 4 & 4 & 1 \\ 3 & 5 & 5 \end{vmatrix} - 5\begin{vmatrix} 1 & 8 & 5 \\ 4 & 4 & 1 \\ 3 & 5 & 5 \end{vmatrix} + 1\begin{vmatrix} 1 & 8 & 5 \\ 2 & 7 & 2 \\ 3 & 5 & 5 \end{vmatrix} - 7\begin{vmatrix} 1 & 8 & 5 \\ 2 & 7 & 2 \\ 4 & 4 & 1 \end{vmatrix}$$

$$= 3 \times (-73) - 5 \times 81 + 1 \times (-62) - 7 \times 53$$

$$= 495.$$

We also have:

$$D_4 = \begin{vmatrix} 3 & 1 & 2 & 8 \\ 5 & 2 & 1 & 7 \\ 1 & 4 & 2 & 4 \\ 7 & 3 & 1 & 5 \end{vmatrix},$$

that is,

$$D_4 = 3\begin{vmatrix} 2 & 1 & 7 \\ 4 & 3 & 4 \\ 3 & 1 & 5 \end{vmatrix} - 5\begin{vmatrix} 1 & 2 & 8 \\ 4 & 3 & 4 \\ 3 & 1 & 5 \end{vmatrix} + 1\begin{vmatrix} 1 & 2 & 8 \\ 2 & 1 & 7 \\ 3 & 1 & 5 \end{vmatrix} - 7\begin{vmatrix} 1 & 2 & 8 \\ 2 & 1 & 7 \\ 4 & 3 & 4 \end{vmatrix}$$

$$= 3 \times (-21) - 5 \times (-45) + 1 \times 12 - 7 \times 39$$

$$= -99.$$

We therefore find:

$$x_1 = \frac{D_1}{|A|} = \frac{198}{99} = 2\,; \qquad x_2 = -\frac{297}{99} = -3\,,$$

$$x_3 = \frac{495}{99} = 5\,; \qquad x_4 = -\frac{99}{99} = -1\,.$$

ANOTHER EXAMPLE

We wish to solve the following system:

$$\begin{aligned}
x_1 - 4x_2 + 3x_3 - 2x_4 &= 7, \\
2x_1 + 3x_2 - 5x_3 + x_4 &= 6, \\
-4x_1 - 6x_2 + 4x_3 + 3x_4 &= 1, \\
x_1 + 7x_2 - 2x_3 - 2x_4 &= -14.
\end{aligned}$$

We have:

$$|A| = \begin{vmatrix} 1 & -4 & 3 & -2 \\ 2 & 3 & -5 & 1 \\ -4 & -6 & 4 & 3 \\ 1 & 7 & -2 & -2 \end{vmatrix} = 0.$$

We verify that all the minors corresponding to the first column are non-zero but that the determinant $|A|$ is itself zero. In this case, the characteristic determinants reduce to a single one since the system is of order 4 and the principal determinant is of order 3.

We then take the principal determinant

$$D = \begin{vmatrix} 1 & -4 & 3 \\ 2 & 3 & -5 \\ -4 & -6 & 4 \end{vmatrix} = -66.$$

We verify that the characteristic determinant D_c is equal to zero, since

$$D_c = \begin{vmatrix} 1 & -4 & 3 & 7 \\ 2 & 3 & -5 & 7 \\ 4 & -6 & 4 & 1 \\ 1 & 7 & -2 & -14 \end{vmatrix} = 0.$$

We then solve the following system:

$$\begin{aligned} x_1 - 4x_2 + 3x_3 &= 7 + 2x_4, \\ 2x_1 + 3x_2 - 5x_3 &= 6 - x_4, \\ -4x_1 - 6x_2 + 4x_3 &= 1 - 3x_4. \end{aligned}$$

We have:

$$D_1 = \begin{vmatrix} 7 & -4 & 3 \\ 6 & 3 & -5 \\ 1 & -6 & 4 \end{vmatrix} - x_4 \begin{vmatrix} -2 & -4 & 3 \\ 1 & 3 & -5 \\ 3 & -6 & 4 \end{vmatrix} = -127 - 67x_4,$$

$$D_2 = \begin{vmatrix} 1 & 7 & 3 \\ 2 & 6 & -5 \\ -4 & 1 & 4 \end{vmatrix} - x_4 \begin{vmatrix} 1 & 2 & 3 \\ 2 & 1 & -5 \\ -4 & 3 & 4 \end{vmatrix} = 191 - 25x_4$$

$$D_3 = \begin{vmatrix} 1 & -4 & 7 \\ 2 & 3 & 6 \\ -4 & 6 & 1 \end{vmatrix} - x_4 \begin{vmatrix} 1 & -4 & -2 \\ 2 & 3 & 1 \\ 4 & -6 & 3 \end{vmatrix} = 143 - 55x_4$$

that is

$$x_1 = \frac{67x_4}{66} + \frac{127}{66}, \qquad x_2 = \frac{25x_4}{66} - \frac{191}{66}, \qquad x_3 = \frac{55x_4}{66} - \frac{143}{66}.$$

CHAPTER VI

BILINEAR FUNCTIONS IN E_n. ORTHOGONALITY. METRIC IN E_n.

So far we have introduced no relationship of "proximity" among the elements of E_n, that is, we have not introduced a topology into the vector space.

Since the concept of proximity is essentially scalar, we will introduce the concept of bilinear functions into vector spaces, associating with every pair of vectors X and Y a scalar function $f(X,Y)$. We will then see how this concept allows us to introduce very interesting new concepts into the study of vector spaces.

(VI-1) *A scalar function associated with any pair of vectors X and Y of E_n is a correspondence associating with every pair (X,Y) a scalar belonging to the field K over which the vector space is defined:*

$$f(X, Y) = a_{xy} \qquad X, Y \in E_n,\ a_{xy} \in K.$$

Except when otherwise stated, we will assume throughout this chapter that K is the field of real numbers.

Note also that our definition of the scalar function is not the most general possible, since the two vectors X and Y belong to the same vector space E_n. We could have assumed that X belonged to E_n and Y to E_m, the only condition imposed on the vector spaces E_n and E_m being that they be defined over the same field K. We give a slightly more restricted definition here, which has the advantage of being simpler and yet adequate for our purposes.

(VI-2) *A scalar function associated with a pair (X,Y) of E_n is said to be bilinear if it satisfies the following conditions:*

(VI-3) $$f(\alpha_1 X_1 + \alpha_2 X_2, Y) = \alpha_1 f(X_1, Y) + \alpha_2 f(X_2, Y),$$

(VI-4) $$f(X, \beta_1 Y_1 + \beta_2 Y_2) = \beta_1 f(X, Y_1) + \beta_2 f(X, Y_2).$$

Hence the following relation:

$$f(\alpha_1 X_1 + \alpha_2 X_2, \beta_1 Y_1 + \beta_2 Y_2) = \alpha_1 \beta_1 f(X_1, Y_1) + \alpha_1 \beta_2 f(X_1, Y_2) + \\ + \alpha_2 \beta_1 f(X_2, Y_1) + \alpha_2 \beta_2 f(X_2, Y_2).$$

We now suppose that the space E_n has a basis $A_1, A_2, \ldots, A_n$; then

$$X = x_1 A_1 + x_2 A_2 + \cdots + x_n A_n$$

$$Y = y_1 A_1 + y_2 A_2 + \cdots + y_n A_n$$

and consequently

$$f(X, Y) = f\left(\sum_i x_i\ A_i, \sum_i y_i\ A_i\right),$$

that is, applying the bilinearity properties of the function f:

$$f(X, Y) = \sum_{i,j} x_i\, y_j\, f(A_i, A_j) = \sum_{i,j} x_i\, y_j\, a_i^j\,,$$

where we have set

$$f(A_i, A_j) = a_i^j\,.$$

We have thus obtained a result that is completely analogous to that obtained in the representation of linear mappings.

(VI-5) *To determine the value of the function f for all possible pairs of vectors X and Y, it is sufficient to know the value of f for all pairs of basis vectors* (A_i, A_j).

Note also that the function $f(X,Y)$ can be expressed in matrix form. We have indeed:

$$f(X, Y) = \sum_i x_i \left(\sum_j a_i^j y_j\right) = X'[M]\, Y,$$

where $[M]$ is the matrix whose elements are a_i^j and X' is the row vector with components $x_1, x_2, \ldots, x_n$.

We have the following result:

(VI-6) *In the space* E_n *with a fixed basis, every bilinear function is represented by a unique matrix* $[M]_{n \times n}$ *; conversely, to a matrix* $[M]_{n \times n}$ *corresponds a unique bilinear form.*

To the bilinear function f there corresponds a unique matrix $[M]$, since it is the matrix whose element in the ith row and jth column is $f(A_i, A_j)$, a a quantity which is uniquely determined. Conversely, the matrix $[M]$ defines all the quantities $f(A_i, A_j)$, which allow us, as we have just seen, to define all the quantities $f(X,Y)$.

We now examine how the bilinear function is affected when we perform a change of basis in E_n.

Let $B_1, B_2, \ldots, B_n$ be a new basis in the space which we previously considered with respect to the basis $A_1, A_2, \ldots, A_n$.

We can write, using the basis $A_1, A_2, \ldots, A_n$:

$$f(X_A, Y_A) = X_A'[M^A]\, Y_A$$

where

$$[M^A] = [f(A_i, A_j)] .$$

Similarly, for the basis $B_1, B_2, \ldots, B_n$:

$$f(X_B, Y_B) = X'_B[M^B]\, Y_B$$

where

$$[M^B] = [f(B_i, B_j)] .$$

And naturally, since the bilinear function is defined independently of the reference system,

$$f(X_A, Y_A) = f(X_B, Y_B)$$

that is,

$$X'_A[M^A]\, Y_A = X'_B[M^B]\, Y_B .$$

But, denoting by $[B]$ the matrix of the change of basis, we know that

$$X_A = [B]\, X_B \qquad\qquad Y_A = [B]\, Y_B .$$

From which we deduce that

$$X'_B[B']\,[M^A]\,[B]\, Y_B = X'_B[M^B]\, Y_B$$

and therefore

(VI-7) $$[M^B] = [B']\,[M^A]\,[B] \quad .$$

We note that this formula is not equivalent to the formula for the transformation of matrices under a change of basis, as stated on page 60. This is natural if we remember that the matrices by no means play the same role in both cases. In the first case, the matrix represents a linear mapping and in the second, a scalar bilinear function. That is why we denote here the change of basis by an upper index for the matrices.

The fact that the form of the matrix $[M]$ changes when the basis is changed presents certain disadvantages in an important case, which we shall consider at a later stage.

There are some changes of basis which reduce these disadvantages: those which are defined by a matrix satisfying the following relation:

(VI-8) $$[B'][B] = [I].$$

(VI-9) *Square matrices satisfying the relation* (VI-8) *are said to be orthogonal and the corresponding changes of basis are called orthogonal changes of basis.*

In view of their importance, we now discuss orthogonal matrices and the corresponding changes of basis at some length.

ORTHOGONAL MATRICES

We have just defined the orthogonal matrices $[C]$ as the set of matrices satisfying the following relation:

$$[C'][C] = [I].$$

We now examine some of their properties.

(VI-10) *An orthogonal matrix is non-singular and its determinant is equal to* 1:

$$[C'][C] = [I].$$

Therefore, looking at the determinants, we get

$$|C'||C| = 1, \text{ that is } |C| = 1.$$

(VI-11) *The inverse of an orthogonal matrix is equal to its transpose.*

This follows from the definition:

$$[C'][C] = [I], \text{ that is } [C'] = [C]^{-1}.$$

(VI-12) *The inverse of an orthogonal matrix is an orthogonal matrix.*

We have indeed

$$[C]^{-1\prime}[C]^{-1} = [C']'[C'] = [C][C'] = [I'] = [I].$$

The matrix $[C']^{-1}$ is therefore orthogonal.

(VI-13) *The product of two orthogonal matrices is an orthogonal matrix.*

If the two matrices $[C_1]$ and $[C_2]$ are orthogonal, we have:

$$[C_1'][C_1] = [I], \qquad [C_2'][C_2] = [I].$$

And we find, considering the product $[C_1][C_2]$:

$$[[C_1][C_2]]'[C_1][C_2] = [C_2'][C_1'][C_1][C_2] = [C_2'][I][C_2]$$
$$= [C_2'][C_2] = [I].$$

The matrix $[C_1][C_2]$ is therefore orthogonal.

(VI-14) *The matrix $[I]$ is orthogonal.*

This is obvious if we remember that $[I]$ is equal to its transpose.

A SPECIAL BILINEAR FUNCTION

Suppose now that given the basis $A_1, A_2, \ldots, A_n$ in E_n, we want the matrix of the bilinear function to be the unit matrix $[I]$

(VI-15) $$f(X, Y) = X'[I]\,Y = X'\,Y.$$

This obviously implies that

(VI-16) $$f(A_i, A_j) = A_i'[I]\,A_j = A_i'\,A_j = 0 \quad i \neq j,$$

(VI-17) $$f(A_i, A_i) = A_i'\,A_i = 1 \qquad i = 1, 2, \ldots, n.$$

This definition of the function f implies very important consequences which we now study in detail.

(VI-18) *The function $f(X,Y)$ is symmetric in X and Y.*

We have indeed

$$f(X, Y) = X'\,Y = (X'\,Y)' = Y'\,X'' = Y'\,X = f(Y, X)\ .$$

The equality of $X'Y$ and its transpose is due to the fact that $X'Y$ is a 1×1 matrix, that is, a scalar.

(VI-19) *$f(X,X)$ is always non-negative.*

We have indeed

$$f(X, X) = X'\,X = x_1^2 + x_2^2 + \cdots + x_n^2$$

which is always positive or zero.

On the other hand, we have just established that the basis vectors $A_1, A_2, \ldots, A_n$ are not arbitrary since they must satisfy the conditions expressed by the relations (VI-16) and (VI-17). Since there are n^2 such conditions and the basis vectors are defined by n^2

parameters, it is always possible to find vectors A_i satisfying these relations. Besides, this can be seen directly by considering the vectors e_i already considered on page 19:

$$\begin{aligned} e_1 &= (1, 0, 0, \dots, 0) \\ e_2 &= (0, 1, 0, \dots, 0) \\ &\vdots \\ e_n &= (0, 0, 0, \dots, 1) . \end{aligned}$$

We have indeed

$$f(X, Y) = X' Y = x_1 y_1 + x_2 y_2 + \cdots + x_n y_n$$

and in particular

$$\begin{aligned} e_i' e_j &= 0 \qquad i \neq j , \\ e_i' e_i &= 1 \qquad i = 1, 2, \dots, n . \end{aligned}$$

(VI-20) *The special bilinear function we have just defined is called the scalar product of the vectors X and Y. It is defined in this form only if the basis vectors of the space satisfy conditions* (VI-16) *and* (VI-17).

(VI-21) *Two vectors X and Y whose scalar product is zero are said to be orthogonal.*

(VI-22) *The square root of the scalar product of a vector X by itself is called the length or norm of the vector X:*

$$\sqrt{X'.X} = \sqrt{x_1^2 + x_2^2 + \cdots + x_n^2} = \| X \| .$$

A vector of length 1 *is said to be normed;* it is a unit vector.

Using these definitions, we can characterize as follows the bases of E_n which allow us to determine the scalar product. They are bases in which the vectors are pairwise orthogonal and of length 1. These bases play such an important role in vector spaces that they have a special name.

(VI-23) *Bases of E_n in which the vectors are pairwise orthogonal and of length* 1 *are called orthonormal bases.*

There is a simple interpretation for the components of a vector X on an orthonormal basis. If $K_1, K_2, \dots, K_n$ are the vectors of such a basis, we have

$$X = x_1 K_1 + x_2 K_2 + \cdots + x_n K_n$$

and

$$K_i' X = \sum_j x_j K_i' K_j = x_i .$$

(VI-24) *The component x_i of a vector X on the vector K_i of an orthonormal basis of E_n is simply the scalar product of K_i by X.*

THE EFFECT OF A CHANGE OF BASIS ON THE SCALAR PRODUCT

We have seen earlier that a change of basis in E_n, transforms the matrix representing the bilinear function as follows:

$$[M^B] = [B'] [M^A] [B] .$$

If the matrix $[M_A]$ represents the scalar product, the above relation becomes

$$[M^B] = [B'] [I] [B] = [B'] [B] ,$$

where the basis $A_1, A_2, \dots, A_n$ is orthonormal.

Consequently, if we want the matrix $[M^B]$ to represent the scalar product in the basis $B_1, B_2, \ldots, B_n$, we must have

$$[M^B] = [B'][B] = [I],$$

or $[B'][B] = [I]$.

This implies that the matrix $[B]$ *is orthogonal.*

Now, in order that the scalar product be meaningful, the basis consisting of the vectors $B_1, B_2, \ldots, B_n$ must be orthonormal. But we have:

$$B_i = [B]^{-1} A_i,$$
$$B_j = [B]^{-1} A_j,$$

and therefore

$$B_i' B_j = A_i' [[B]^{-1}]' [B]^{-1} A_j = A_i' [[B][B]']^{-1} A_j = A_i' A_j = 0$$

and similarly

$$B_i' B_i = 1.$$

Hence, in this case, the basis $B_1, B_2, \ldots, B_n$ is orthonormal as well as the basis $A_1, A_2, \ldots, A_n$.

This is a particular case of the following result:

(VI-25) *A change of basis with an orthogonal matrix preserves scalar product, orthogonality and length of vectors.*

The scalar product is preserved since

$$X_A' Y_A = X_B'[B'][B]Y_B = X_B' Y_B;$$

for the matrix $[B]$, being orthogonal, satisfies the relation

$$[B]'\,[B] = [I]\,.$$

Since orthogonality and length are expressed in terms of the scalar product alone, they are invariant under a change of basis with orthogonal matrix.

We have mentioned earlier (VI-14) that the matrix $[I]$ whose column vectors are the vectors $e_1, e_2, \ldots, e_n$ is orthogonal. This is a particular case of the following theorem:

(VI-26) *An $n \times n$ matrix $[C]$ is orthogonal if and only if its column vectors form an orthonormal basis of the space E_n, where the basis used to represent the matrix is itself orthonormal.*

The condition is necessary.

Since the matrix $[C]$ is orthogonal, it satisfies the following relation:

$$[C']\,[C] = [I]\,.$$

which implies the following relations between the elements of the matrix $[C]$:

$$\sum_l c_l^{\,i}\, c_l^j = \begin{cases} 0 & \text{if } i \neq j \\ 1 & \text{if } i = j \end{cases}. \qquad (1)$$

Now, these relations express the fact that the column vectors are pairwise orthogonal and have unit length.

The condition is sufficient.

If the column vectors are orthogonal, they satisfy the relations (1), which can be expressed as follows:

$$\sum_l (c_i^l)'\,(c_l^j) = \begin{cases} 0 & \text{if } i \neq j\,, \\ 1 & \text{if } i = j\,, \end{cases}$$

or

$$[C'][C] = [I].$$

The matrix $[C]$ is therefore orthogonal, Q.E.D.

The converse of Theorem (VI-25) can be deduced from the above result:

The matrix of a change of basis by which one goes from one orthonormal basis to another orthonormal basis is orthogonal.

We know indeed that the column vectors of this matrix represent the components on the vectors of the first basis of the vectors of the second basis; since the latter form an orthonormal system, it follows from the previous theorem that the corresponding matrix is orthogonal.

It follows from all this that we must use orthonormal bases of E_n whenever we want to introduce the concept of a scalar product, that is, of a metric. We shall also see that these concepts allow us to introduce the concept of distance very simply and naturally into vector spaces.

Whenever we consider vector spaces consisting of sets of n scalars $(a_1, a_2, \ldots, a_n)$ in this order, it is very easy to find an orthonormal basis, the set of n vectors $e_1, e_2, \ldots, e_n$. We obtain other orthonormal bases from this basis by performing changes of basis defined by orthogonal matrices.

In any case, it is very important, given a basis of E_n, to be able to transform it into an orthonormal basis. This is called *orthonormalization* of the bases of E_n.

ORTHONORMALIZATION OF THE BASES OF E_n

This operation of orthonormalization requires the use of a process we are now going to define.

Orthogonal projection of a vector X of E_n onto a vector V of E_n

The projection of a vector X of E_n onto another vector V of E_n is a vector collinear with V, defined as follows:

$$\text{Proj}_V X = \frac{X'V \,.\, \sqrt{X'X}}{\sqrt{X'X \,.\, V'V}} \frac{V}{\sqrt{V'V}},$$

that is,

$$\text{Proj}_V X = \frac{X'V}{V'V} V.$$

The projection reduces to the zero vector θ if the vectors X and V are orthogonal.

It is readily verified that the first form we gave for the projection is merely a simple generalization of the concept of projection in ordinary two-dimensional Euclidean space, derived from the scalar product.

If we consider the scalar product of the vector V by the vector $X - \text{Proj}_V X$, we find:

$$V'(X - \text{Proj}_V X) = V'X - \frac{X'V}{V'V} V'V = V'X - X'V = 0.$$

In other words, the vector V and the vector $X - \text{Proj}_V X$ are orthogonal. We immediately make use of this property in the orthonormalization of bases.

Let $A_1, A_2, \ldots, A_n$ be a basis of E_n. We are going to derive a new orthonormal basis $B_1, B_2, \ldots, B_n$ of E_n from this first basis, proceeding as follows: Let us take

$$B_1 = \frac{A_1}{\sqrt{A_1' A_1}},$$

which implies $\| B_1 \| = 1$.

We then take

$$B_2 = \frac{A_2 - (B_1' A_2) B_1}{| A_2 - B_1' A_2 B_1 |} = \frac{A_2 - B_1' A_2 A_1/\sqrt{A_1' A_1}}{| A_2 - B_1' A_2 B_1 |}$$

which implies

$$B_1' B_2 = 0 \quad \text{and} \quad \| B_2 \| = 1 .$$

We also take

$$B_3 = \frac{A_3 - (B_1') A_3 B_1 - (B_2' A_3) B_2}{| A_3 - (B_1') A_3 B_1 - (B_2) A_3 B_2 |},$$

which implies

$$B_1' B_3 = 0 , \qquad B_2' B_3 = 0 , \qquad \| B_3 \| = 1, \text{etc} .$$

In other words, at each stage of the simplification process we take as the vector B_i a vector deduced from A_i by subtracting its projections on $B_1, B_2, ..., B_{i-1}$ and dividing the vector obtained in this way by its length. We obtain each time a vector B_i of length 1 which is orthogonal to each of the vectors $B_1, B_2, ..., B_{i-1}$. When we reach the vector B_n, we have obtained an orthonormal basis. During these various operations, we are sure B_i will never be the zero vector. In fact, we can readily verify step by step that the vector B_i can be expressed as follows:

$$B_i = \alpha_2^1 A_1 + \alpha_2^2 A_2 + \cdots + \alpha_i^{i-1} A_{i-1}$$

where the α_i^j are not all identically zero. It follows that B_i can be equal to the vector θ only if the vectors $A_1, A_2, ..., A_{n-1}$ are linearly dependent, which is impossible since the set $A_1, A_2, ..., A_n$ is a basis of E_n.

We see that the matrix $[A]$ of the change of basis is not arbitrary. It is of the following form:

$$[A] = \begin{bmatrix} a_1^1 & a_1^2 & \dots & a_1^n \\ 0 & a_2^2 & \dots & a_2^n \\ \vdots & & & \\ 0 & 0 & \dots & a_n^n \end{bmatrix}.$$

All elements below the main diagonal are zero. Although its column vectors form an orthonormal basis of E_n, the matrix $[A]$ is not orthogonal, which seems to contradict Theorem (VI-26). This can be explained by the fact that the basis used to represent the matrix $[A]$ is not orthonormal (the A_i are not orthonormal). It follows that the theorem does not apply in this case.

The orthonormalization process we have just described, which is due to Erhard Schmidt, allows us to state the following result:

(VI-29) *Given a basis* $A_1, A_2, \dots, A_n$ *of* E_n *it is always possible to deduce an orthonormal basis* $B_1, B_2, \dots, B_n$.

As pointed out earlier in this chapter, we shall always assume that the space E_n is referred to an orthonormal basis.

The scalar product allows us to give a very simple definition of a metric in E_n.

METRIC IN E_n

In the vector space E_n defined over the field K of real numbers, we construct a distance as follows:

The distance between two vectors X and Y of E_n is a scalar function $d(X,Y)$ of these two vectors which satisfies the following conditions:

(VI-30) $d(X, X) = 0$

(VI-31) $d(X, Y) = d(Y, X)$ (commutativity)

(VI-32) $d(X, X) \geqslant 0$

(VI-33) If $d(X, Y) = 0$, $X = Y$
(VI-34) $d(X, Z) \leqslant d(X, Y) + d(Y, Z)$ (triangle inequality)

These conditions are not sufficient to define a distance since several types of functions satisfy them. We therefore make use of the scalar product and the length defined earlier in order to obtain a distance in E_n.

We first point out some useful properties of length, after recalling its definition:

(VI-35) *The length of a vector X is the square root of the scalar product (X,X) - X'X and is generally written as follows:*

$$\| X \| = \sqrt{(X, X)} = \sqrt{X' X} = \sqrt{x_1^2 + x_2^2 + \cdots + x_n^2}.$$

The latter form immediately yields two important properties of length:

(VI-36) *The length of a vector is a non-negative number.*

(VI-37) *The vector θ alone has length zero.*

The length of a vector also has the following properties:

$$\| aX \| = a \| X \|, \qquad a \in R, \qquad X \in E_n,$$

since

$$\| aX \| = \sqrt{(aX, aX)} = \sqrt{a^2 X' X} = a \sqrt{X' X} = a \| X \|.$$

GENERALIZED THEOREM OF PYTHAGORAS

(VI-38) *The square of the length of the sum of two vectors X and Y of E_n is equal to the sum of the squares of their lengths if and only if the vectors are orthogonal.*

We have indeed

$$\| X + Y\|^2 = (X + Y, X + Y) = X'X + 2X'Y + Y'Y = \| X \|^2 + \| Y\|^2 + 2X'Y$$

and the scalar product $X'Y$ is zero if and only if the two vectors X and Y are orthogonal.

DISTANCE BETWEEN TWO VECTORS OF E_n

On page 123 we introduced the concept of distance between two vectors, indicating the properties such a function must have. We now clarify this concept as follows:

(VI-39) *In a space E_n, the distance between two vectors X and Y is the length of the vector $X - Y$, that is,*

$$d(X, Y) = \| X - Y\| = \sqrt{(X - Y, X - Y)},$$

or

$$d(X, Y) = \sqrt{X'X + Y'Y - 2X'Y}.$$

It is readily verified that the function $d(X,Y)$ introduced above satisfies the properties (VI-30) through (VI-33) on page 123. We now show that it also satisfies property (VI-34).

Since distance is a positive scalar (Property (VI-32)), condition (VI-34) can also be expressed as follows:

$$d^2(X, Z) \leqslant d^2(X, Y) + d^2(Y, Z) + 2\,d(X, Y)\,d(Y, Z),$$

that is,

$$\| X - Z\|^2 \leqslant \| X - Y\|^2 + \| Y - Z\|^2 + 2\sqrt{(X - Y, X - Y)(Y - Z, Y - Z)}.$$

But on the other hand, we can also write

$$\| X - Z\|^2 = \| X - Y + Y - Z\|^2 = (X - Y + Y - Z, X - Y + Y - Z)$$
$$= \| X - Y\|^2 + \| Y - Z\|^2 + 2(X - Y, Y - Z).$$

We can therefore write the above inequality as follows:

$$\| X - Y \|^2 + \| Y - Z \|^2 + 2(X - Y, Y - Z) \leqslant \| X - Y \|^2 + \| Y - Z \|^2 + 2\sqrt{(X - Y, X - Y)(Y - Z, Y - Z)}\,.$$

In other words, we must prove the inequality:

$$\text{(VI-40)} \qquad (X - Y, Y - Z) \leqslant \sqrt{(X - Y, X - Y)(Y - Z, Y - Z)}\,,$$

which can be written as follows:

$$(U, V) \leqslant \sqrt{(U, U)(V, V)}\,.$$

Consider the expression $\| U + tV \|^2$, where t is a scalar. It can be written as follows:

$$\| U + tV \|^2 = (U + tV, U + tV) = (U, U) + 2\,t(U, V) + t^2(V, V)\,.$$

This expression - a quadratic form in t - is always positive (since it is a square); it follows that its discriminant is negative and that

$$(U, V)^2 - (V, V)(U, U) \leqslant 0\,.$$

Hence the inequality (VI-40) - called the Schwartz inequality - and therefore also the inequality we wished to prove.

Later, in our study of the decomposition of vector spaces, we shall meet an interesting application of the concept of distance.

CHAPTER VII

DIAGONALIZATION OF SQUARE MATRICES OF SIMPLE STRUCTURE

We have seen, in Chapter VI, the importance of orthogonal matrices and orthogonal changes of basis. On the other hand, on page 66 (Chapter III), we met diagonal matrices and scalar matrices and we saw how useful they are in many cases. This will be apparent again in the following chapters.

When possible, it is very useful to perform orthogonal changes of basis in the vector spaces under consideration, as they allow us to transform matrices of linear mappings into diagonal matrices while preserving length and distance. Finding such a change of basis and the new form of a square matrix $[A]_{n\times n}$ is called *diagonalizing* this matrix. We are going to see that this is possible only for certain types of matrices, which are said to have simple structure.

We shall always assume in the sequel that the space E_n is defined over the field K of complex numbers and that it is initially referred to an orthonormal basis, say the basis consisting of the vectors e_1, e_2, ..., e_n.

We first introduce the concept of an eigenvector of a matrix with n rows and n columns.

(VII-1) *A vector Λ of E_n is said to be an eigenvector of the matrix $[F]$ if*

$$[F]\,\Lambda = k\Lambda \qquad k \in K\,.$$

(VII-2) *The scalar k is called the eigenvalue associated with the eigenvector Λ.*

(VII-3) *If a vector* Λ *is an eigenvector of the matrix* $[F]$*, so is every vector* $\alpha\Lambda$*, where* α *is a non-zero scalar.*

We have indeed, from the properties of linear mappings:

$$[F]\,\alpha\Lambda = \alpha\,[F]\,\Lambda = \alpha k\Lambda\,.$$

$\alpha\Lambda$ is therefore an eigenvector, and the eigenvalue associated with it is again k.

In the sequel, we shall always consider eigenvectors of length 1.

The relation defining an eigenvector,

$$[F]\,\Lambda = k\Lambda\,,$$

can be written as follows:

$$[F]\,\Lambda - k\Lambda = \theta\,,$$

or, using the additive property of linear mappings,

$$\big[[F] - k[I]\big]\,\Lambda = \theta\,.$$

(VII-3) *If the vector* X *is not the zero vector, the relation* $[G]X = \theta$ *is possible if and only if the determinant of the matrix* $[G]$ *is zero.*

The condition is necessary.

The vector $[G]X$ is of form

$$[G]\,X = x_1[G]\,e_1 + x_2[G]\,e_2 + \dots + x_n[G]\,e_n\,.$$

It follows that the relation $[G]X = \theta$ will be satisfied without all the x_i being identically zero only if the vectors $[G]e_i$ are linearly dependent. The determinant of their matrix $[G]$ is therefore equal to zero.

The condition is sufficient.

If the determinant of $[G]$ is zero, there is at least one linear relation between its column vectors $[G]\,e_1, [G]\,e_2, \ldots, [G]\,E_n$; consequently, if we write down the system of n equations representing the coordinates of the vector $[G]X$, we know that this system reduces to $n - r$ equations if $[G]$ is of rank r. Theorem (I-31) can therefore be applied, since we are dealing with a system of $n - r$ equations in n unknowns; this system admits at least one non-trivial solution and there is at least one vector X, not identically zero, which satisfies the relation $[G]X = \theta$.

Hence, in the case of an eigenvector, we must have

$$|\,[F] - k[I]\,| = 0\,.$$

Now the determinant $|\,[F] - k[I]\,|$ is of the following form:

$$|\,[F] - k[I]\,| = \begin{vmatrix} f_1^1 - k & f_1^2 & \cdots & f_1^n \\ f_2^1 & f_2^2 - k & \cdots & f_2^n \\ \cdot & \cdot & \cdot & \cdot \\ f_n^1 & f_n^2 & \cdots & f_n^n - k \end{vmatrix} = D(k)\,.$$

Expanding it, we see that it is a polynomial of degree n in k: *it is the characteristic polynomial of the matrix* $[F]$ *and the equation* $|\,[F] - k[I]\,| = 0$ *is the characteristic equation of the matrix* $[F]$.

This equation always admits n roots since the scalars f_i^j belong to the field of complex numbers; these roots are the *eigenvalues* of the matrix $[F]$.

Two cases may occur:

(VII-4a) *The matrix* $[F]$ *is non-singular.*

Then

The eigenvalues are all different from zero.

If there were an eigenvalue equal to zero, we should have

$$|[F] - k[I]| = |F| = 0,$$

which is impossible since the matrix $[F]$ is non-singular.

(VII-5b) *The matrix* $[F]$ *is of rank* r.

Then

Zero is a root of the characteristic equation.

This is clear since the determinant $|F|$ is equal to zero. It can even easily be shown by using Schur's Theorem (VII-10) (below page 134), that 0 is a root of multiplicity $n - r$ when the matrix $[F]$ is symmetric.

Let Λ_i be an eigenvector associated with the eigenvalue k_i.

(VII-6) *Vectors* $\Lambda_1, \Lambda_2, \ldots, \Lambda_j$ *corresponding to distinct roots* $k_1, k_2, \ldots, k_j$ *of the characteristic equation are linearly independent.*

Suppose they are not. This would mean that there is a zero linear combination of these vectors whose coefficients are not all zero and that we could, for instance, express Λ_j in terms of $\Lambda_1, \Lambda_2, \ldots, \Lambda_j$, that is

$$\Lambda_j = \lambda_1^j \Lambda_1 + \lambda_2^j \Lambda_2 + \cdots + \lambda_{j-1}^j \Lambda_{j-1}.$$

If we apply the matrix $[F]$ to the two sides of this relation, we find

$$[F] \Lambda_j = \sum_i \lambda_i^j [F] \Lambda_i.$$

Since the vectors Λ_i are eigenvectors of the linear mapping with matrix $[F]$, we have

$$[F]\,\Lambda_i = k_i\,\Lambda_i\,,$$

and the relation above can be written as follows:

$$k_j\,\Lambda_j = \sum \lambda_i^j\,k_i\,\Lambda_i\,.$$

Replacing then, in the left-hand side, Λ_j by its value in terms of the Λ_i and identifying the coefficients of $\Lambda_1, \Lambda_2, ..., \Lambda_{j-1}$ in both sides, we find the following relations:

$$k_j\,\lambda_i^j = k_i\,\lambda_i^j \qquad \text{for} \qquad i = 1, 2, ..., j-1\,.$$

And these relations are impossible since the k_i are all different by hypothesis.

(VII-7) *If the roots of the characteristic equation of a non-singular matrix* $[F]_{n\times n}$ *are all distinct, there exists a change of basis which will transform the matrix* $[F]$ *into a diagonal matrix.*

Indeed, if the roots of the characteristic equation are all distinct, the preceding theorem shows that there exist n linearly independent eigenvectors $\Lambda_1, \Lambda_2, ..., \Lambda_n$. These vectors therefore form a basis of E_n. If we perform a change of basis, taking $\Lambda_1, \Lambda_2, ..., \Lambda_n$ as the new basis, the matrix $[F]$ will take the following form (where $[\Lambda]$ is the matrix of the change of basis):

$$[F_\Lambda] = [\Lambda]^{-1}\,[F]\,[\Lambda] = \begin{vmatrix} k_1 & 0 & \dots & 0 \\ 0 & k_2 & \dots & 0 \\ \vdots & & & \\ 0 & 0 & \dots & k_n \end{vmatrix}\,.$$

In fact, since each column vector C_i is the transform of the basis vector A_i, it is equal to $k_i A_i$. We see that the matrix $[F_A]$ is non-singular, since its determinant - equal to $k_1 k_2 \dots k_n$ - is non-zero, which is to be expected since the matrices $[\Lambda]$ and $[A]^{-1}$ are non-singular.

The characteristic equation of the matrix $[F]$ has the following interesting property:

(VII-8) *The characteristic equation of a matrix $[F]$ is invariant under a change of basis in the space E_n.*

This follows immediately from Theorem (IV-26) since the characteristic polynomial of $[F]$ is simply the determinant of the matrix $[F] - k[I]$.

In the case where the characteristic equation has multiple roots, the results we have just established are no longer valid in general, and are replaced by far more complex results which we shall study in Chapter VIII.

We nevertheless establish now a general result which will lead us to examine the case of symmetric matrices in more detail.

The theorem we have in mind calls for a few preliminary results.

(VII-9) *Given an orthonormal basis in the space E_n and a vector C_1 of length 1, we can always complete the latter by $n - 1$ other vectors $C_2, \dots, C_n$ such that the matrix with the vectors $C_1, C_2, \dots, C_n$ as its column vectors is orthogonal.*

Indeed, given a vector C_1 of E_n, we know that we can always complete it by $n - 1$ vectors $A_2, \dots, A_n$ such that the vectors $C_1, A_2, \dots, A_n$ form a basis of E_n. But we know, by Schmidt's method, that given the n vectors

C_1, $C_2, \ldots, C_n$, it is possible to deduce an orthonormal basis of E_n, which we call $C_1, C_2, \ldots, C_n$ (since C_1 is of length 1). It follows that the matrix with the orthonormal vectors C_i as its column vectors is an orthogonal matrix.

(VII-10) *If the matrix* $[C_n]$ *with n rows and n columns is orthogonal, the same is true of the matrix* $[C_{n+1}]$ *defined as follows:*

$$[C_{n+1}] = \begin{bmatrix} 1 & 0 & 0 & \ldots & 0 \\ 0 & . & . & . & . \\ \vdots & & & & \\ \vdots & . & . & C_n & . \\ \vdots & & & & \\ 0 & . & . & . & . \end{bmatrix}.$$

It is sufficient to verify that the column vectors of $[C_{n+1}]$ form an orthonormal basis of E_{n+1}. Now we find, denoting by C_0^* the first column vector of $[C_{n+1}]$ and by C_j^* its jth column vector:

$$\sum_0^n c_i^0 c_i^j = 0, \qquad j = 1, 2, \ldots, n$$

$$| C_0^* | = 1$$

and

$$\sum_0^n c_i^j c_i^k = c_0^j c_0^k + \sum_1^n c_i^j c_i^k = 0 + 0,$$

$$| C_j^* |^2 = \sum_0^n c_2^{j2} = 0 + \sum_1^n c_2^{j2} = 1.$$

The vectors C_j^* are therefore orthonormal and the matrix of which they form the column vectors is orthogonal.

We could also have affirmed directly that

$$[C'_{n+1}][C_{n+1}] = [I]$$

since

$$[C'_{n+1}][C_{n+1}] = \begin{bmatrix} 1 & 0 & \cdots\cdots & 0 \\ 0 & \cdot & \cdots & \cdot \\ \vdots & & & \\ \vdots & \cdot & [C'_n]\,[C_n] & \cdot \\ 0 & \cdot & \cdots & \cdot \end{bmatrix} = [I]$$

as we can see by calculating the elements of $[C'_{n+1}][C_{n+1}]$ directly.

We are now in position to state the following important theorem:

Schur's Theorem.

(VII-11) *Given a matrix $[A]_{n\times n}$ with real, though not necessarily distinct, eigenvalues, there is an orthogonal change of basis which transforms the matrix A into a matrix B in which all elements below the principal diagonal are zero (triangular matrix).*

We prove this theorem by induction, showing that if it is assumed true for an $n \times n$ matrix, it is also true for an $(n + 1) \times (n + 1)$ matrix.

We first show that it is true for a 2×2 matrix. Let $[A]_{2\times 2}$ be a matrix whose eigenvalues are real.

Let Λ_1 be the eigenvector of this matrix associated with an eigenvalue λ_1. We know that it is possible to find a vector M_2 such that the two vectors Λ_1 and M_2 form an orthonormal basis of E_2. This follows from the fact that in a space E_n it is always possible to complete p linearly independent vectors (in this case $p = 1$) with $n - p$ vectors such that the set obtained in this way forms a basis of E_n and also from Schmidt's method for orthonormalizing this basis (Theorems (I-35) and (VI-29)).

We therefore perform a change of basis in E_2, taking the vectors Λ_1 and M_2 as the new basis. The matrix $[C]$ of the change of basis is then orthogonal (we are always assuming that the original basis is orthonormal).

In the new basis, the matrix $[A]$ will take the following form:

$$[A_c'] = [C]^{-1}\,[A]\,[C] = \begin{bmatrix} \lambda^1 & b_1^1 \\ 0 & b_2^1 \end{bmatrix}$$

since

$$[A]\,\Lambda_1 = \lambda_1\,\Lambda_1\,.$$

We deduce from this representation that b_2^1 is the second eigenvalue of the matrix $[A]$.

Indeed, since the characteristic equation remains invariant under the change of basis, we can write:

$$(\lambda - \lambda_1)(\lambda - \lambda_2) = (\lambda - \lambda_1)(\lambda - b_2^1)\,,$$

whence

$$b_2^1 = \lambda_2\,.$$

b_2^1 may be equal to λ_1, and we have

$$[A_C] = [C]^{-1}\,[A]\,[C] = [C']\,[A]\,[C] = \begin{bmatrix} \lambda_1 & b_1^1 \\ 0 & \lambda_2 \end{bmatrix}.$$

The result is therefore true for a matrix with two rows and two columns.

We now assume that it is true for n and prove that it is also true for $n + 1$.

Consider a matrix $[A]$ with $n + 1$ rows and $n + 1$ columns.

Let Λ_1 be an eigenvector of this matrix associated with an eigenvalue λ_1.

It is possible to find an orthogonal matrix $[C]_{(n+1) \times (n+1)}$ whose first column is Λ_1.

By means of the change of basis defined by $[C]$, we transform the matrix $[A]$ into $[A_C]$, where

$$[A_C] = [C'] [A] [C] = \begin{vmatrix} \lambda_1 & b_1^1 & b_1^2 & \dots & b_1^n \\ 0 & b_2^1 & b_2^2 & \dots & b_2^n \\ 0 & \dots & \dots & \dots & \dots \\ 0 & b_{n+1}^1 & b_{n+1}^2 & \dots & b_{n+1}^n \end{vmatrix} .$$

We set

$$[B] = \begin{vmatrix} b_2^1 & b_2^2 & & b_2^n \\ b_3^1 & b_3^2 & & b_3^n \\ \dots & \dots & \dots & \dots \\ b_{n+1}^1 & b_{n+1}^2 & & b_{n+1}^n \end{vmatrix} .$$

The matrix $[B]$ has n rows and n columns; the theorem therefore applies to it, and there exists an orthogonal matrix $[C_n]$ such that

$$[B_{C_n}] = [C_n'] [B] [C_n] = \begin{vmatrix} \mu_2 & \beta_2^3 & & \beta_2^{n+1} \\ 0 & \mu_3 & & \beta_3^{n+1} \\ \vdots & & & \\ 0 & 0 & \dots\ 0 & \mu_{n+1} \end{vmatrix} .$$

But, on the other hand, the characteristic polynomial of the matrix $[A_C]$, which is the same as that of the matrix $[A]$, is

$$| A_C - \lambda I | = \begin{vmatrix} \lambda_1 - \lambda & b_1^1 & b_1^2 & \dots & b_1^n \\ 0 & b_2^1 - \lambda & b_2^2 & \dots & b_2^n \\ \dots & \dots & \dots & \dots & \dots \\ 0 & b_{n+1}^1 & b_{n+1}^2 & \dots & b_{n+1}^{n-\lambda} \end{vmatrix} .$$

Expanding with respect to the elements in the first column:

$$| [A_C] - \lambda I | = (\lambda_1 - \lambda) \begin{vmatrix} b_2^1 - \lambda & b_2^2 & \dots & b_2^n \\ b_3^1 & b_3^2 - \lambda & \dots & b_3^n \\ \dots & \dots & \dots & \dots \\ b_{n+1}^1 & b_{n+1}^2 & \dots & b_{n+1}^n - \lambda \end{vmatrix}$$

or

$$| A_C - \lambda I | = (\lambda_1 - \lambda) \, | [B] - \lambda[I] | .$$

In other words, these n eigenvalues of the matrix $[B]$ are merely the other eigenvalues of $[A]$, namely $\lambda_2, \lambda_3, \dots, \quad \lambda_{n+1}$.

Since the eigenvalues of $[B_{C_n}]$ are the same, we can write:

$$[B_{C_n}] = [C_n'] \, [B] \, [C_n] = \begin{vmatrix} \lambda_2 & \beta_2^3 & & \beta_2^{n+1} \\ 0 & \lambda_3 & & \beta_3^{n+1} \\ \vdots & & & \\ 0 & 0 & \dots & \lambda_{n+1} \end{vmatrix} .$$

Consider then a matrix $[C_{n+1}]$ derived from $[C_n]$ by adding a first row (1,0,0,...,0) and a first column (k,0,0,...,0):

$$[C_{n+1}] = \begin{bmatrix} 1 & 0 & 0 & \dots & 0 \\ 0 & & & & \\ 0 & & & & \\ \vdots & & C_n & & \\ \vdots & & & & \\ 0 & & & & \end{bmatrix}.$$

By Theorem (VII-10), $[C_{n+1}]$ is an orthogonal matrix since $[C_n]$ is orthogonal.

$$[A_C][C_{n+1}] = \begin{bmatrix} \lambda_1 & \sum_{i=1}^{n} b_1^i C_i^1 & \dots & \sum_{i=1}^{n} b_1^i C_i^n \\ 0 & \sum_{i=1}^{n} b_2^i C_i^1 & \dots & \sum_{i=1}^{n} b_2^i C_i^n \\ \vdots & & & \\ 0 & \sum_{i=1}^{n} b_{n+1}^i C_i^1 & \dots & \sum_{i=1}^{n} b_{n+1}^i C_i^n \end{bmatrix}$$

which can be written as follows

$$[A_C][C_{n+1}] = \begin{bmatrix} \lambda_1 & \sum_{i=1}^{n} b_1^i C_i^1 & \dots & \sum_{i=1}^{n} b_1^i C_i^n \\ 0 & & & \\ \vdots & & & \\ \vdots & & [B][C_n] & \\ 0 & & & \end{bmatrix}.$$

And we have seen earlier that the matrix $[C_n'][B][C_n]$ is triangular. It follows therefore that the matrix $[C_{n+1}'][A_C][C_{n+1}]$ is also triangular:

$$[C'_{n+1}]\,[A_C]\,[C_{n+1}] = \begin{bmatrix} \lambda_1 & \sum_{i=1}^{n} b_1^i C_i^1 & \cdots\cdots & \sum_{i=1}^{n} b_1^i C_i^n \\ 0 & & & \\ \vdots & & [C'_n]\,[B]\,[C_n] & \\ \vdots & & & \\ 0 & & & \end{bmatrix}.$$

But on the other hand:

$$[C'_{n+1}]\,[A_C]\,[C_{n+1}] = [C'_{n+1}]\,[C']\,[A]\,[C]\,[C_{n+1}]$$

$$= [CC_{n+1}]'\,[A]\,[CC_{n+1}] = [K']\,[A]\,[K].$$

and the matrix $[K]$, the product of two orthogonal matrices, is itself orthogonal.

We have thus found an orthogonal matrix $[K]$ which transforms the matrix $[A]$ into a triangular matrix.

The theorem has been proved in the case where the eigenvalues of $[A]$ are real, but it can also be proved in the identical way for the case where $[A]$ has complex eigenvalues. It is sufficient to replace everywhere the term orthogonal matrix by that of unitary matrix.

APPLICATION TO SYMMETRIC MATRICES

Symmetric matrices play a very important role in the theory of quadratic forms and the results we have just established in the general case become particularly interesting in the case of symmetric matrices.

(VII-12) *The eigenvalues of a symmetric matrix are all real.*

So far, in Chapters V and VI, we have given definitions of transpose matrices and of scalar products which are valid when the field used to define the vector spaces is the field R of real numbers. In other words, the elements under consideration were real. It is clear that the roots of the characteristic equation may be complex. We therefore give here the most general definition of transpose matrices and of scalar product, when the elements defining the vector space belong to the field K of complex numbers.

If the matrix $[A]$ has general element a_i^k, the matrix $[A]'$ has the general element $\bar{a}_k^i$, where the symbol $\bar{a}$ denotes the complex conjugate of a, that is,

$$a \cdot \bar{a} = \bar{a} \cdot a = |a|^2 = r^2 .$$

Under these conditions, the scalar product of two vectors X and Y of E_n can be expressed as follows:

$$X' Y = \sum \bar{x}_i y_i = \sum \bar{y}_i x_i .$$

Note that the definitions above are the same as the definitions given earlier in the case where the elements belong to the field of real numbers.

Consider the symmetric matrix $[F]$ and the following scalar product, where X is a vector of E_n:

$$X'[F] X = \sum_{i,k} f_i^k \bar{x}_i x_k .$$

We also have

$$\overline{X'[F] X} = \sum_{i,k} \bar{f}_k^i \bar{\bar{x}}_i \bar{x}_k .$$

Since the matrix $[F]$ is real and symmetric, we have

$$\overline{f_i^k} = f_i^k = f_k^i$$

and on the other hand

$$\bar{\bar{x}}_i = x_i .$$

Hence

$$\overline{X' [F] X} = \sum_{i,k} f_k^i x_i \bar{x}_k = X'[F] X$$

for any X. It follows that the scalar product $X'[F]X$, which is always equal to its conjugate, is real for any real complex X.

We then consider the following expression in which Λ_i is an eigenvector:

$$\Lambda_i'[F] \Lambda_i = \Lambda_i' k_i \Lambda_i = k_i \Lambda_i' \Lambda_i = k_i ,$$

since the eigenvector Λ_i has length 1. It follows that k_i is real.

(VII-13) *If all the eigenvalues of a symmetric matrix are distinct, the eigenvectors* $\Lambda_1, \Lambda_2, \ldots, \Lambda_n$ *form an orthonormal basis of* E_n.

We have seen earlier (VII-7) that the n vectors $\Lambda_1, \Lambda_2, \ldots, \Lambda_n$ form a basis of E_n. To show that this basis is orthonormal, let us consider the scalar product of the vectors $[F] \Lambda_i$ and $[F] \Lambda_j$. We have:

$$f([F] \Lambda_i, [F] \Lambda_j) = \Lambda_j'[F'] [F] \Lambda_i .$$

But the matrix $[F]$ is symmetric, so that

$$[F'] = [F] .$$

This implies

$$\Lambda_j'[F_j'] [F] \Lambda_i = \Lambda_j' [F]^2 \Lambda_i = \Lambda_j' k_i^2 \Lambda_i = k_i^2 \Lambda_j' \Lambda_i .$$

But we have directly

$$f([F] \Lambda_i, [F] \Lambda_j) = f(k_i \Lambda_i, k_j \Lambda_j) = k_i k_j \Lambda_j' \Lambda_i$$

which implies

$$k_i k_j A'_j A_i = k_i^2 A'_j A_i .$$

Since $k_i \neq k'_j$, this relation is possible only if

$$A'_j A_i = 0 .$$

The vectors A_i are therefore pairwise orthogonal. As they also have length 1, they form an orthonormal basis of E_n.

Thus the change of basis which diagonalizes the matrix $[F]$ is orthogonal (we are still assuming that the initial basis is orthonormal).

We can now extend this result to the case where the eigenvalues of the symmetric matrix $[A]$ are no longer distinct.

Schur's Theorem shows that there always exists an orthogonal change of basis which transforms the matrix $[A]$ into a triangular matrix. In other words, there is an orthogonal matrix $[K]$ such that the matrix

$$[A_k] = [K'] [A] [K]$$

is triangular.

If we transpose the matrix $[A_k]$, we find

$$[A_k]' = ([K]' [A] [K])' = [K]'([K]' [A])' = [K]' [A]' [K]'' = [K]' [A] [K] .$$

In other words, the matrix $[A_k]$ is also symmetric, like the matrix $[A]$. Since it is also triangular because of Schur's Theorem, it is a diagonal matrix. We have thus established the following general result:

(VII-14) *Given an arbitrary symmetric matrix $[A]$ with real elements, there is an orthogonal matrix $[K]$ such that the matrix $K'[A]K$ is diagonal.*

This result shows that all the eigenvectors Λ_i of the matrix $[A]$ are pairwise orthogonal, even those corresponding to the same eigenvalue [sic].

Indeed, after the change of basis defined by the matrix $[K]$, the column vectors of $[A_k]$ are the transforms of the new basis vectors under the linear mapping with matrix $[A_k]$. If we denote by $L_1, L_2, \ldots, L_n$ the new basis vectors of E_n (they are orthonormal since the matrix $[K]$ of the change of basis is orthogonal), we find

$$A_k^i = [A_k]\, L_i = k_i\, L_i\,.$$

The vectors L_i are therefore the eignevectors of the matrix $[A_k]$. They are also the eigenvectors of the matrix $[A]$ since eigenvectors are invariant under an orthogonal change of basis. We see this by writing

$$[A]\, \Lambda_i = k_i\, \Lambda_i$$

and then expressing Λ_i in the new basis. We have indeed, denoting by Λ_k^i the representation of Λ_i in the new basis:

$$\Lambda_i = [K]\, \Lambda_k^i\,.$$

We can therefore write

$$[A]\, \Lambda_i = k_i\, \Lambda_i = [A]\,[K]\, \Lambda_k^i = k_i\,[K]\, \Lambda_k^i\,.$$

Applying $[K]^{-1}$ to both sides, we find

$$[K]^{-1}\,[A]\,[K]\, \Lambda_k^i = k_i[K]^{-1}\,[K]\, \Lambda_k^i = k_i\, \Lambda_k^i\,,$$

or

$$[A_k]\, \Lambda_k^i = k_i\, \Lambda_k^i\,,$$

which shows that the vector Λ_k^i is an eigenvector of the matrix $[A_k]$.

We have thus shown that the eigenvectors of the symmetric matrix $[A]$ form an orthonormal basis of E_n if the matrix $[A]$ is non-singular. If it is of rank p, we find that 0 is an eigenvalue with multiplicity $n - p$.

CHAPTER VIII

DECOMPOSITION OF VECTOR SPACES INTO SUBSPACES. APPLICATION TO THE REDUCTION OF SQUARE MATRICES

We have already defined (Chapter I) the concept of a subspace of a space E_n and mentioned some of its properties. In this chapter, we consider some subspaces with special properties.

We begin with a few definitions and properties.

(VIII-I) *The space E_n is called the direct sum of the subspaces E_{n1} and E_{n2} if every vector V of E_n can be decomposed into a sum $V_1 + V_2$, where V_1 is in E_{n1} and V_2 is in E_{n2}, and the two subspaces have only the zero vector θ in common.*

The space E_n is also said to be decomposed into two subspaces E_{n1} and E_{n2}.

In other words, every vector V of E_n can be expressed as follows:

$$V = V_1 + V_2\,, \qquad V_1 \in E_{n_1}\,, \qquad V_2 \in E_{n_2}\,.$$

Several properties follow readily from the above definitions.

(VIII-2) *The decomposition of a vector V of E_n is unique.*

Suppose we had found two different decompositions. We would have:

$$V = V_1 + V_2\,, \qquad V_1 \in E_{n_1}\,, \qquad V_2 \in E_{n_2}\,,$$

$$V = V_1' + V_2'\,, \qquad V_1' \in E_{n_1}\,, \qquad V_2' \in E_{n_2}\,,$$

that is, we could write

$$V_1 + V_2 = V_1' + V_2'$$

or

$$V_1 - V_1' = V_2 - V_2'\,, \qquad V_1 - V_1' \in E_{n_1}\,, \qquad V_2' - V_2 \in E_{n_2}\,.$$

Now, by hypothesis, the two subspaces E_{n1} and E_{n2} only have the zero vector θ in common; it follows that

$$V_1 - V_1' = \theta\,, \qquad V_2 - V_2' = \theta\,,$$

which proves the uniqueness of the representation.

(VIII-3) *The sum of the dimensions of the subspaces E_{n1} and E_{n2} is equal to the dimension of the space E_n.*

Consider indeed a vector X of E_n. It can be expressed as follows:

$$X = X' + X''\,, \qquad X' \in E_{n_1}\,, \qquad X'' \in E_{n_2}$$

where

$$X' = \sum_1^{n_1} x_i' A_i'\,, \qquad X'' = \sum_1^{n_2} x_j'' A_j''\,,$$

the A_i' forming a basis of E_{n1}, the A_j'' a basis of E_{n2}.

We can write

$$X = \sum_i^{n_1} x_i' A_i' + \sum_j^{n_2} x_j'' A_j''\,.$$

Now, the vectors A_i' are independent, and so are

the vectors A''_j, and the vectors A'_i and A''_j considered as a whole are also independent. If this were not the case, at least one vector A'_i would be a linear combination of the vectors A''_j and would hence belong to E_{n2}, which is impossible since θ is the only vector common to the subspaces E_{n_1} and E_{n2}.

It follows that every vector X in E_n is a linear combination of $n_1 + n_2$ linearly independent vectors A'_i and A''_j, which implies that n is equal to $n_1 + n_2$.

SUBSPACES INVARIANT UNDER A LINEAR MAPPING OF E_n INTO ITSELF

Consider a subspace E_p of E_n. Given a linear mapping of E_n into itself, we say that

(VII-4) *The subspace E_p is invariant under this linear mapping if every vector of E_p is transformed by the linear mapping into a vector of E_p.*

The whole space E_n is invariant under any linear mapping of E_n into itself.

The importance of the concept of invariance of a subspace is that it allows us to study a linear mapping of E_n into itself in each invariant subspace, since the behavior of this mapping in the subspace in question is completely independent of its behavior in the rest of the space E_n.

The decomposition of a space into special subspaces allows us to display some interesting properties of this space or of the linear mappings that can be defined, particularly when considering invariant subspaces.

DECOMPOSITION OF E_n INTO ORTHOGONAL SUBSPACES

We consider only the decomposition of E_n into two subspaces, since generalization to more than two subspaces is immediate.

Consider a p-dimensional subspace of E_n, E_p say, where $p < n$ of course.

(VII-5) *A vector Z of E_n is said to be orthogonal to E_p if it is orthogonal to all the vectors in E_p, that is, if*

$$U' Z = 0$$

for any $U \in E_p$.

(VIII-6) *A vector Z of E_n is orthogonal to E_p if and only if it is orthogonal to all the vectors of a basis of E_p.*

The condition is obviously necessary. If Z is orthogonal to all the vectors of E_p, it is in particular orthogonal to the p basis vectors $A_1, A_2, \ldots, A_p$.

The condition is sufficient. Every vector X of E_p is of the form

$$X = \sum x_i A_i$$

and therefore

$$X' Z = \sum x_i A_i' Z = 0 .$$

(VIII-7) ***Definition.*** *The projection of a vector X of E_n onto a subspace E_p of E_n is the vector formed by the sum of the projections of this vector X onto the vectors of an orthogonal basis of E_p. This projection U is unique no matter what orthogonal basis of E_p we are considering.*

To show the uniqueness of the definition, let us

consider an orthogonal basis of the subspace E_p consisting of the vectors $A_1, A_2, \ldots, A_p$. We have

$$U = \sum_i \frac{A_i' X}{A_i' A_i} A_i .$$

We also consider another basis of E_p consisting of the vectors $B_1, B_2, \ldots, B_p$. The projection of X onto the vectors B_i is given by

$$V = \sum_i \frac{B_i' X}{B_i' B_i} B_i .$$

If we now consider the vectors $X - U$ and $X - V$, we see that they are orthogonal to E_p, since

$$A_i'(X - U) = A_i' X - A_i' U = A_i' X - \sum_k \frac{A_k' X}{A_k' A_k} A_k' A_i ,$$

$$B_j'(X - V) = B_j' X - B_j' V = B_j' X - \sum_k \frac{B_k' X}{B_k' B_k} B_k' B_j .$$

Now all the terms in $A_k' A_i$ with $k \neq i$, and in $B_k' B_j$ with $k \neq j$, are zero because of the orthogonality of the bases, whence

$$A_i'(X - U) = A_i' X - \frac{A_i' X}{A_i' A_i} A_i' A_i = 0$$

and similarly

$$B_j'(X - V) = 0 .$$

Since the vectors $X - U$ and $X - V$ are orthogonal to E_p, so is their difference $(X - U) - (X - V)$, which can also be written as

$$X - U - X' + V = V - U.$$

It follows that the vector $V - U$, which is both in E_p and orthogonal to E_p, must be the vector θ, and we have $U = V$.

We have also proved in passing that every vector X of E_n can be expressed as a sum of two vectors Y and Z, where Y belongs to a subspace E_p of E_n, and Z is orthogonal to E_p. *The set of vectors Z defined in this way forms a vector subspace E of E_n.* To show this, it is sufficient to show that the zero vector θ is in this set and that the operations of addition and multiplication by a scalar are closed in it. Now, the zero vector θ belongs to this set. On the other hand, if we consider two vectors Z_1 and Z_2 of E and an arbitrary vector X of E_p, we have

$$X'(Z_1 + Z_2) = X' Z_1 + X' Z_2 = 0 + 0 = 0 .$$

It follows that $Z_1 + Z_2$ is orthogonal to every X of E_p and therefore belongs to E. Similarly, if Z belongs to E, if α is a scalar and if X is a vector of E_p, we have

$$X'(\alpha Z) = \alpha X' Z = 0 ,$$

and αZ therefore belongs to E_p.

Every vector X in E_p can therefore be decomposed into the sum of two vectors Y and Z belonging respectively to E_p and to E. In other words, X can be decomposed into two orthogonal vectors Y and Z.

We can therefore say that

(VIII-8) *The space E_n is the direct sum of the two orthogonal subspaces E_p and E.*

It follows that the subspace E has dimension $n - p$.

This follows directly from Theorem (VIII-3).

We now state and prove an important theorem:

(VIII-9) *Given a vector space E_n, a subspace E_p and a vector X of E_n which is not in E_p, the vector of E_p at the shortest distance from X is the projection of X onto E_p.*

Consider a vector A of E_p and let P_x be the projection of X onto E_p. Then

$$A = P_x + B$$

and the square of the distance between X and A can be written

$$\| X - A \|^2 = (X - A)'(X - A) = (X - P_x - B)'(X - P_x - B)$$
$$= (X - P_x)'(X - P_x) - B'(X - P_x) - (X - P_x)'B + B'B .$$

Since $X - P_x$ is orthogonal to B, this relation reduces to

$$\| X - A \|^2 = \| X - P_x \|^2 + \| B \|^2 ,$$

and this expression is minimal for

$$\| B \| = 0 ,$$

that is,

$$A = P_x .$$

DECOMPOSITION OF E_n INTO CYCLIC SUBSPACES

This very important type of decomposition leads us to the definition of some concepts of algebraic nature concerning linear mappings.

Minimal polynomial of a vector of E_n with respect to a linear mapping of E_n into itself.

Consider in E_n an arbitrary vector X and a linear mapping of E_n into itself represented by a matrix $[M]_{n \times n}$.

We can associate the vectors $[M]\, X$, $[M]^2\, X$, ..., $[M]^n\, X$, ... etc, with the vector X. Since we are dealing with an n-dimensional space, we can find at most n such linearly independent vectors. Suppose there are p of them. If we consider the sequence

$$X, [M]\,X, [M]^2\,X, ..., [M]^p\,X$$

the $p + 1$ vectors which form it are linearly dependent and one of them, say $[M]^p\,X$, can be expressed as a linear combination with coefficients not all zero of the vectors $[M]^k\,X$. For instance, assigning the coefficient +1 to the term in $[M]^p\,X$ yields

$$[M]^p\,X + \alpha_1[M]^{p-1}\,X + \cdots + \alpha_p\,X = \theta\,,$$

which can be written as follows

$$\varphi(M)\,X = \theta\,.$$

(VIII-10) *The polynomial $\varphi(z)$ is then said to be a singular polynomial for the vector X with respect to the linear mapping with matrix $[M]$.*

In the rest of this chapter we will omit the reference to the linear mapping with matrix $[M]$ since we will always consider the same one.

(VIII-11) *The polynomial $\varphi(z)$ defined above is the polynomial of smallest degree satisfying the relation $\varphi(M)X = \theta$. It is the minimal polynomial of X.*

If we had been able to find a polynomial $\varphi'(z)$ degree smaller than that of φ and satisfying the relation

$$\varphi'(M)\,X = \theta\,,$$

this would have meant that we had found less than $p + 1$ vectors of the form $[M]^k X$ which are connected by a linear relation with coefficients not all zero, that is, less than p linearly independent vectors $[M]^k X$, which contradicts our hypothesis.

(VIII-12) *Every singular polynomial for the vector X is a multiple of the minimal polynomial $\varphi(z)$.*

Consider a singular polynomial φ' of X; it can be written as follows:

$$\varphi'(z) = \varphi(z)\,h(z) + r(z)\,,$$

$r(z)$, the remainder on division of φ' by φ, is of smaller degree than φ. We have thus

$$\varphi'(M)\,X = \varphi(M)\,h(M)\,X + r(M)\,X = \theta + r(M)\,X = \theta\,,$$

since φ' and φ are singular polynomials for X. It follows that

$$r(M)\,X = \theta\,.$$

Hence $r(z)$ is a singular polynomial of X, of degree smaller than p, which is impossible. It follows that $r(z)$ is identically zero and that φ' is indeed a multiple of φ.

Moreover, every polynomial which is a multiple of the minimal polynomial of X is a singular polynomial, which follows immediately from the commutativity of multiplication of polynomials.

N.B. Since the minimal polynomial of a vector X is always defined up to a scalar, we agree to consider always the polynomial $\varphi(z)$ in which the leading term, i.e., the term in z^p, has coefficient 1.

(VIII-13) *The minimal polynomial* $\varphi(z)$ *of a vector* X *is unique.*

Suppose we had found two minimal polynomials $\varphi(z)$ and $\varphi'(z)$; we would obviously have

$$\varphi(M)\,X = \theta$$

and

$$\varphi'(M)\,X = \theta\,.$$

And also

$$[\varphi(M) - \varphi'(M)]\,X = \theta\,.$$

The polynomial $\varphi(M) - \varphi'(M)$ would then be a singular polynomial for X; now, it is a polynomial of degree at most $p - 1$ since the coefficient of z^p in φ and in φ' is equal to 1; we would thus have found a singular polynomial for X of degree smaller than p, which contradicts the hypothesis that φ and φ' are minimal polynomials, hence

$$\varphi(z) \equiv \varphi'(z)\,.$$

(VIII-14) *The minimal polynomial of the sum of k vectors* $A_1, A_2, \ldots, A_k$ *whose minimal polynomials are relatively prime is equal to the product of these minimal polynomials.*

We prove this in the case of two vectors A_1 and A_2, since it is sufficient to repeat the reasoning in order to obtain the general case.

Let $\varphi_1(z)$ and $\varphi_2(z)$ be the respective minimal polynomials of A_1 and A_2. It is obvious that the polynomial $\varphi_1(z)\varphi_2(z)$ is a singular polynomial for the sum $A_1 + A_2$. We have indeed

$$\varphi_1[M]\,\varphi_2[M]\,(A_1 + A_2) = \varphi_1[M]\,\varphi_2[M]\,A_1 + \varphi_1[M]\,\varphi_2[M]\,A_2 = \theta\,.$$

But on the other hand, if we consider a singular polynomial $\varphi(z)$ of the sum $A = A_1 + A_2$, we find

$$\begin{aligned}\varphi_1[M]\,\varphi[M]\,A_2 &= \varphi_1[M]\,\varphi[M]\,(A - A_1)\\ &= \varphi_1[M]\,\varphi[M]\,A - \varphi_1[M]\,\varphi[M]\,A_1 = \theta - \theta\,.\end{aligned}$$

$\varphi_1(z)\varphi(z)$ is therefore a singular polynomial of A_2 and is consequently divisible by $\varphi_2(z)$. But since φ_1 and φ_2 are relatively prime, this implies that φ is divisible by $\varphi_2(z)$. We could also show in the same way that $\varphi(z)$ is divisible by $\varphi_1(z)$. Hence $\varphi(z)$ must be of form $\varphi_1(z)\varphi_2(z)\psi(z)$ and the minimal polynomial of A is therefore $\varphi_1(z)\varphi_2(z)$.

We can extend the theorem by stages to the case of more than two vectors A_1 and A_2.

We now consider a basis of E_n consisting of n vectors $A_1, A_2, \ldots, A_n$ and let $\varphi_1(z), \varphi_2(z), \ldots, \varphi_n(z)$ be their respective minimal polynomials. *The product* $p(z) = \Pi\varphi_i(z)$ *is a singular polynomial for any vector X of* E_n. Indeed, every vector X of E_n can be expressed as follows:

$$X = \sum_i^n x_i\,A_i\,,$$

which implies

$$p(M)\,X = \sum_{i}^{n} x_i\, p(M)\, A_i$$

and since $p(z)$ is singular for each A_1 we have

$$p(M)\, A_i = \theta$$

and

$$p(M)\, X = \theta\,.$$

(VIII-15) *The polynomial $p(z)$ is then said to be a singular polynomial for the space E_n, that is, we have in fact*

$$p(M) = [0]\,.$$

It is readily shown that the polynomial $p(z)$ is the singular polynomial of E_n with smallest degree, that is, the minimal polynomial of E_n. This minimal polynomial must indeed be minimal for all the vectors of a basis of E_n and must therefore be a common multiple of the minimal polynomials of these basis vectors, and it must be the least common multiple. It is therefore precisely the polynomial $p(M)$ we have just defined.

The polynomial $p(z)$ has been defined in terms of a special basis $A_1, A_2, \ldots, A_n$. But we have the following result:

(VIII-16) *The minimal polynomial of a space E_n is unique.*

Suppose indeed that we had found two minimal polynomials $p(z)$ and $p'(z)$, we would have

$$p(M) = [0]\,,$$
$$p'(M) = [0]\,,$$
$$p(M) - p'(M) = [0]\,.$$

The polynomial $p(z) - p'(z)$ would therefore be a singular polynomial for E_n; now, since the leading terms of p and p' have the same coefficient +1, it follows that the degree of $p - p'$ is smaller by one than that of p and p', which contradicts the hypothesis that p and p' are minimal. We have thus

$$p(z) \equiv p'(z).$$

We are now going to state a very fundamental result.

(VIII-17) *If the minimal polynomial of a space E_n is the product of two relatively prime polynomials, the space can be decomposed into the direct sum of two invariant subspaces having respectively the two component polynomials as their minimal polynomials.*

Suppose that the minimal polynomial of E_n can be expressed as follows:

$$p(z) = p_1(z)\, p_2(z),$$

where p_1 and p_2 are relatively prime.

Consider in E_n the vectors which admit respectively $p_1(z)$ and $p_2(z)$ as their singular polynomials. These two sets of vectors form two subspaces E_{p_1} and E_{p_2} of E_n. We then show, for instance, that the sum of two vectors of E_{p_1} is a vector of E_{p_1}; consider two vectors X_1 and X_1' of E_{p1}. We have

$$p_1(M)\,(X_1 + X_1') = p_1(M)\,X_1 + p_1(M)\,X_1'$$

that is, since X_1 and X_1' are in E_{p1},

$$p_1(M)\,(X_1 + X_1') = \theta + \theta = \theta.$$

$X_1 + X_1'$ therefore belongs to E_{p_1}. We could prove in a similar fashion the other properties showing that the two sets E_{p_1} and E_{p_2} are vector subspaces of E_n.

Consider now a vector X of E_n, and write it in the following form:

$$X = X_1 + X_2, \text{ where } X_1 \in E_{p_1}, X_2 \notin E_{p_1}.$$

X can always be expressed in this form; it is sufficient, for instance, to take the projection of X onto E_{p_1} as X_1. We then have

$$\begin{aligned} p(M)X &= p(M)(X_1 + X_2) = p(M)X_1 + p(M)X_2 \\ &= p_1(M)p_2(M)X_1 + p_1(M)p_2(M)X_2 = p_1(M)p_2(M)X_2 = \theta, \end{aligned}$$

since

$$p_1(M)p_2(M)X_1 = \theta.$$

It follows that $p_1(z)p_2(z)$ is a singular polynomial of X_2, and since $p_1(z)$ is not a singular polynomial of X_2 (since p_1 and p_2 are relatively prime), $p_2(z)$ must be a singular polynomial for X_2. Therefore X_2 is in E_{p_2}. We have decomposed X into two vectors, the first of which is in E_{p_1} and the second in E_{p_2}.

To complete the proof, we must show that a vector X cannot be in E_{p_1} and in E_{p_2} unless it is the vector θ. Indeed, if X were in both E_{p_1} and E_{p_2}, we would have

$$p_1(A)X = \theta, \qquad p_2(A)X = \theta.$$

p_1 and p_2 would then be singular polynomials for X, which would imply that they have common factors, which is impossible since they are relatively prime.

E_n has therefore been decomposed into the direct sum of two subspaces which, by their very construction, admit p_1 and p_2, respectively, as their minimal polynomials.

N.B. The result we have just proved of course remains valid if the polynomial $p(z)$ is decomposed into a product of k polynomials which are all relatively prime. The space E_n then decomposes into the direct sum of k invariant spaces $E_{p_1}, E_{p_2}, \ldots, E_{p_K}$. This can readily be proved step by step.

Now we can always express a polynomial $p(z)$ as the product of integer powers of irreducible polynomials,

$$p(z) = p_1(z)^{k_1}\, p_2(z)^{k_2} \ldots p_r(z)^{k_1}.$$

If the scalar field is the field of complex numbers, we know that a polynomial of degree n always has exactly n roots and the minimal polynomial can therefore be written as follows:

$$p(z) = (z - z_1)^{k_1}(z - z_2)^{k_2} \ldots (z - z_r)^{k_r}$$

where $\sum_1^r k_i$ = degree of the minimal polynomial and the corresponding space E_n decomposes into the direct sum of r invariant vector subspaces with respectively $(z - z_1)^{k_1}, (z - z_2)^{k_2}, \ldots, (z - z_r)^{k_r}$ as minimal polynomials. We shall examine this question again at a later stage.

CYCLIC SUBSPACES

We associated earlier to a vector X of E_n and to a linear mapping of E_n into E_n the sequence of vectors:

$$X, [M]\,X, [M]^2\,X, \ldots, [M]^n\,X, \ldots$$

where $[M]$ is the matrix of the linear mapping under consideration. Let p be the maximum number of these linearly independent vectors.

(VIII-18) *The vectors* $X, [M]\,X, \ldots, [M]^{p-1}\,X$, *define a p-dimensional vector subspace of* E_n. *This space, which is invariant under the linear mapping with matrix* $[M]$, *is said to be cyclic; it is generated by the vector* X.

It follows from Theorem (I-34) that the vectors $X, [M]X, \ldots, [M]^{p-1}X$, define a vector space E_p .

The term "cyclic" derives from the following:

If we set

$$X = A_1, [M]X = A_2, \ldots, [M]^k X = A_{k+1}, \ldots, [M]^{p-1} X = A_p,$$

we see, from the very manner in which we defined the A_i , that

$$[M]A_1 = A_2, [M]A_2 = A_3, \ldots, [M]A_k = A_{k+1}, \ldots,$$

The space is invariant, since the $p + 1$ vectors $X, [M]X, \ldots, [M]^p X$ are linearly dependent, which allows us to write:

$$[M]^p X = [M]A_p = \sum_1^p \alpha_i A_i .$$

Now a vector of E_p can be expressed as follows:

$$Y = \sum_1^p y_j A_j$$

and consequently

$$[M]Y = \sum_1^p y_j [M]A_j = \sum_1^{p-1} A_{j+1} + y_p[M]A_p,$$

that is,

$$[M]Y = \sum_1^{p-1} y_j A_{j+1} + y_p \sum_1^p \alpha_i A_i = \sum_1^p y'_j A_j .$$

This shows that $[M]Y$ is in E_p, and hence that this subspace is invariant under the linear mapping we are discussing.

Note the following result:

(VIII-19) *The minimal polynomial of the subspace* E_p *is the minimal polynomial of the vector* X *which generated this subspace.*

We must show that the following relation is valid for every vector Y of E_p :

$$\varphi(M)\, Y = \theta\,,$$

which can be written as follows:

$$\sum_1^p y'_j\, \varphi[M]\, [M]^{j-1}\, X = \theta\,.$$

For if $\varphi(z)$ is the minimal polynomial of X, it is obvious that $\varphi(z)\, z^{j-1}$ is a singular polynomial for X for all values of j. It follows that all the terms in the sum above are equal to θ, so that their sum is also θ, and there is no polynomial $\psi(z)$ of degree smaller than that of $\varphi(z)$ for which $\psi(M)Y$ is equal to θ for any $Y \in E_p$. To see this we need only take X itself as our Y.

This is also a special case of the following theorem:

(VIII-20) *In every space* E_n *there is a vector which admits the minimal polynomial of the space as its minimal polynomial.*

To prove this theorem, we make use of the remark on page 159. The minimal polynomial of the space E_n can be written as follows:

$$p(z) = [p_1(z)]^{k_1}\, [p_2(z)]^{k_2} \dots [p_r(z)]^{k_r}$$

where the p_i are relatively prime and the k_i are integers.

Suppose that $p(z)$ reduces to the following form:

$$p(z) = [p_1(z)]^{k_1}.$$

Consider a basis $A_1, A_2, ..., A_n.$ in E_n .

We know that $p(z)$ is the smallest common multiple of the minimal polynomials of the A_i. It follows that all these minimal polynomials are of the form $[p_i(z)]^{k_i}$ and that $p(z)$ is equal to the minimal polynomial of the vector A_i with the largest exponent.

In the general case where $p(z)$ has more than one irreducible factor, we shall use Theorem (VIII-14).

The space E_n has been decomposed into the direct sum of the spaces $E_{p_1}, E_{p_2}, ..., E_{p_r}$, admitting respectively

$$[p_1(z)]^{k_1}, [p_1(z)]^{k_2}, ..., [p_r(z)]^{k_r}$$

as their minimal polynomials. Hence, in each subspace E_{p_i}, we can find a vector A_i whose minimal polynomial is the minimal polynomial of the subspace. From Theorem (VIII-14), the vector

$$A = A_1 + A_2 + \cdots + A_r$$

admits the product of the minimal polynomials of its terms - that is, $p(z)$ - as its minimal polynomial.

(VIII-21) *An n-dimensional vector space E_n is cyclic if and only if its minimal polynomial has degree n.*

The condition is necessary. If the space E_n is cyclic, it contains a basis $A_1, A_2, ..., A_n$ whose vectors satisfy the following conditions:

$$A_i = [M]\, A_{i-1},$$

which implies

$$A_2 = [M]\,A_1\,, \qquad A_3 = [M]^2\,A_1, \ldots, A_n = [M]^{n-1}\,A_1\,.$$

And since any $n + 1$ vectors of E_n are necessarily dependent, it follows that

$$[M]^n\,A_1 = \sum \alpha_k [M]^k\,A_1\,,$$

or

$$\sum_1^n \alpha_k [M]^k\,A_1 = \varphi(M)A_1 = \theta$$

and $\varphi(z)$ is the minimal polynomial of A_1.

If there were a polynomial $\varphi'(z)$ of degree m smaller than n such that $\varphi'(M)\,A_1$ is the zero vector θ, this would mean that the vectors $A_1, A_2, \ldots, A_n$ are not linearly independent, which contradicts the hypothesis that they form a basis of E_n. Moreover, we know that if $\varphi(z)$ is the minimal polynomial of A_1, it is also the minimal polynomial of the vector space generated by A_1, and hence of E_n.

The condition is sufficient. $\varphi(z)$, the minimal polynomial of E_n, is of degree n. By Theorem (VIII-20), there exists a vector A of E_n with $\varphi(z)$ as its minimal polynomial. This implies that the n vectors $A, [M]\,A, \ldots, [M]^{n+1}\,A$ are linearly independent. They thus form a basis of E_n and satisfy the condition:

$$A \;= [M]\,A_{k-1}\,.$$

The space E_n is therefore cyclic.

This result yields a decomposition of the space E_n into cyclic subspaces.

(VIII-22) *An n-dimensional vector space E_n can always be decomposed into a direct sum of invariant cyclic subspaces.*

We have just seen that if the minimal polynomial of E_n is of degree n, the space E_n is cyclic. The theorem is therefore true in that case.

Suppose that the minimal polynomial $p(z)$ of E_n is of degree smaller than n. As previously, we can find a vector A in E_n with $p(z)$ as its minimal polynomial. The p vectors $A, [M]A, \ldots, [M]^{p-1}$ are therefore linearly independent and they generate a cyclic vector subspace E_p of E_n ; since $p < n$, E_p is a proper subspace of E_n and we know that every vector X of E_n can be expressed as follows:

$$X = Y + Z$$

where $Y \in E_p$ and $Z \notin E_p$.

The vectors Z defined in this way form an $(n - p)$-dimensional vector subspace E_{n-p} of E_n and E_n is the direct sum of the subspaces E_p and E_{n-p}.

The subspace E_{n-p} has a minimal polynomial of degree r, $R(z)$, and there is a vector B in E_{n-p} such that $R(z)$ is its minimal polynomial. It follows that the vectors $B, [M]B, \ldots, [M]^{r-1}B$ are linearly independent and that they generate a subspace of E_n, E_r say, which is cyclic and invariant. We also show that given a vector $Z \in E_r$, the vector $[M]^k Z$ is in E_{n-p} for any k. Suppose indeed that this is not the case. Then

$$[M]^k Z = Y_k + Z_k, \qquad Y_k \in E_p, Z_k \in E_{n-p}.$$

If we apply the operator $R(z)$ to both sides, we get

$$R(M)[M]^k Z = R(M) Y_k + R(M) Z_k.$$

The left-hand side and the second term in the right-hand side are zero, since Z and Z_k are in E_{n-p}, whose minimal polynomial is $R(z)$; it follows that

$$R(M)\, Y_k = \theta\,.$$

Now, since the space E_p is invariant, $R[M]\, Y_k$ is in E_p. Therefore

$$R(M)\, Y_k = \sum_{0}^{p-1} \alpha_j [M]^j A = F(M)\, A = \theta\,,$$

where the degree of $F(z)$ is at most equal to $p - 1$. We have thus found a singular polynomial of A with degree smaller than p, which contradicts the hypothesis that the minimal polynomial of A has degree p. Hence

$$Y_k = \theta\,.$$

It follows that the vectors $B, [M]\, B, \ldots, [M]^{r-1} B$ are linearly independent of the vectors $A, [M]\, A, \ldots, [M]^{p-1} A$, so that the set of vectors $[M]\, A, \ldots, [M]^{p-1} A, B, [M]\, B, \ldots, [M]^{r-1} B$ forms a basis for the vector space consisting of the direct sum of the spaces E_p and E_r. If $p + r = n$, we have decomposed the space E_n into the direct sum of two cyclic spaces; if not, we continue the decomposition process in the same way. Since E_n is finite-dimensional, a finite number of operations will yield a sequence of subspaces whose direct sum is equal to E_n.

Note moreover that since $R(z)$ is the minimal polynomial of E_{n-r} it divides the minimal polynomial of E_n, that is

$$P(z) = Q(z)\, R(z)\,.$$

And on the other hand, $R(z)$ is also the minimal polynomial of E_r. In general, the minimal polynomial of the space obtained at any stage of the decomposition divides the minimal polynomial of the space obtained at the previous stage.

We can carry the decomposition even further:

Consider the first subspace E_p we obtained above. Its minimal polynomial $P(z)$ can be decomposed into a product of irreducible factors:

$$P(z) = P_1(z)^{k_1} P_2(z)^{k_2} \dots P_m(z)^{k_m}.$$

It follows from Theorem (VIII-22) that the subspace E_r can be decomposed into a sum of m invariant subspaces whose minimal polynomials are respectively $P_1(z)^{k_1}, \dots, P_m(z)^{k_m}$.

The polynomials $P(z)$, $R(z)$ etc. which we defined earlier can be decomposed in a similar fashion, so that the space E_n is decomposed into invariant subspaces with integer powers of irreducible polynomials as their minimal polynomials. Moreover, these invariant subspaces are also cyclic, like the initial subspaces E_p, E_r, etc.

This follows from a theorem we will now prove:

(VIII-23) *A cyclic space can only be decomposed into cyclic subspaces with relatively prime polynomials.*

Consider the cyclic space E_n which decomposes into two invariant subspaces E_r and E_t. Let $P(z)$, $R(z)$ and $T(z)$ be the minimal polynomials of E_n, E_r and E_t respectively; they are respectively of degree m, m_r and m_t. The corresponding spaces have dimension n, r and t, where $n = r + t$.

The polynomial $P(z)$ is the least common multiple of the polynomials $R(z)$ and $T(z)$, so that

$$m \leqslant m_r + m_t .$$

But on the other hand, we must have

$$m_r \leqslant r , \qquad m_t \leqslant t ,$$

that is,

$$m_r + m_t \leqslant r + t$$

or

$$m \leqslant m_r + m_t \leqslant n .$$

But $m = n$, since E_n is cyclic, and therefore

$$m_r + m_t = n ,$$

which implies the following relations

$$m_r = r , \qquad m_t = t , \qquad m_r + m_t = m .$$

Since m_r is equal to r and m_t is equal to t, the spaces E_r and E_t are cyclic. Since m is the sum of m_r and m_t, the polynomials $R(z)$ and $T(z)$ are relatively prime.

We now use these results to transform square $n \times n$ matrices in the cases that have not been examined in Chapter VII.

CANONICAL FORM OF A MATRIX. JORDAN CANONICAL FORM

Consider a cyclic space whose polynomial is $R(z)$, where

$$R(z) = z^n + a_1 z^{n-1} + a_2 z^{n-2} + \cdots + a_n .$$

Consider in E_n a vector A with minimal polynomial $R(z)$ and let us use the following vectors as a basis of E_n :

$$\begin{aligned} A_1 &= A, \\ A_2 &= [M]\ A = [M]\, A_1, \\ A_3 &= [M]^2\, A = [M]\, A_2, \\ &\cdots\cdots\cdots\cdots \\ A_n &= [M]^{n-1}\, A = [M]\, A_n. \end{aligned}$$

The matrix $[M]$ has the following form in this basis:

$$[M] = \begin{vmatrix} 0\ 0 & \dots & -a_n \\ 1\ 0 & \dots & -a_{n-1} \\ 0\ 1 & & \\ \vdots\ \vdots & & \\ 0\ 0 & & 1 - a_1 \end{vmatrix}$$

since

$$\begin{aligned} [M]\, A_n = [M]^n\, A &= -\, a_1[M]^{n-1}\, A - a_2[M]^{n-2}\, A - \dots - a_n\, A \\ &= -\, a_1\, A_n - a_2\, A_{n-1} - \dots - a_n\, A. \end{aligned}$$

If we consider the characteristic equation of the matrix $[M]$, namely

$$|\,[M] - \lambda[I]\,| = 0,$$

we find

$$|\,[M] - \lambda[I]\,| = \begin{vmatrix} -\lambda & 0 & \dots & & -a_n \\ 1 & -\lambda & \dots & & -a_{n-1} \\ \vdots & & & & \\ 0 & 0 & \dots & 1 & -a_1 - \lambda \end{vmatrix}$$

$$= -\, a_n - \lambda\, a_{n-1} - \lambda^2\, a_{n-2} - \cdots + \lambda^{n-1}(-\, a_1 - \lambda) = R\,(\lambda).$$

We have thus proved the following result:

(VIII-24) *The characteristic polynomial and minimal polynomial of a cyclic vector space* E_n *are identical*

Consider now a non-cyclic space E_n ; we have seen that it can be decomposed into a direct sum of cyclic subspaces $E_p, E_r, \ldots, E_v$.

If we take the vectors $A, [M]A, \ldots, [M]^{p-1}A, B, [M]B, \ldots, [M]^{r-1}B, \ldots, G, [M]G, \ldots, [M]^{v-1}G$ as a basis of E_n, the matrix $[M]$ can be written in the form (α) given on the next page.

The blocks containing only zeros occur because the subspaces E_j are invariant, so that the images of the basis vectors $D_1, D_2 = [M]D$, etc., can only be expressed in terms of these vectors themselves. We can write the matrix as follows:

$$[M] = \begin{vmatrix} [M_p] & & & 0 \\ & [M_r] & & \\ & & \ddots & \\ 0 & & & [M_v] \end{vmatrix} .$$

Each matrix $[M_j]$ is characterized by the fact that its minimal polynomial is equal to its characteristic polynomial, and the characteristic polynomial of the matrix $[M_j]$ is divisible by the characteristic polynomial of the matrix $[M_{j+1}]$.

Moreover, when writing the characteristic equation of the matrix $[M]$, we see that

$$|\,[M] - \lambda[I]\,| = P(\lambda)\,R(\lambda) \ldots V(\lambda) ,$$

$$
[M] = \begin{array}{c} p \left\{ \begin{array}{c} \\ \\ \\ \\ \\ \end{array} \right. \\ r \left\{ \begin{array}{c} \\ \\ \\ \\ \end{array} \right. \\ \\ \\ v \left\{ \begin{array}{c} \\ \\ \\ \\ \\ \end{array} \right. \end{array}
\left|
\begin{array}{c:c:c:c}
\overbrace{\begin{array}{cccccc}
0 & 0 & 0 & \ldots\ldots\ldots & & -a_p \\
1 & 0 & 0 & \ldots\ldots\ldots & & -a_{p-1} \\
0 & 1 & 0 & \ldots\ldots\ldots & & -a_{p-2} \\
\ldots & \ldots & \ldots & \ldots\ldots\ldots & & \\
0 & 0 & 0 & \ldots\ldots\ldots & 1 & -a_1
\end{array}}^{p}
&
\overbrace{\begin{array}{ccc}
0 & 0 & \ldots\ldots\ldots\ldots\ldots\ldots \\
 & & \\
 & 0 & \\
 & & \\
 & &
\end{array}}^{r}
&
\begin{array}{c} 0 \\ \\ 0 \\ \\ \\ \end{array}
&
\overbrace{\begin{array}{c} \\ \\ 0 \\ \\ \\ \end{array}}^{v}
\\ \hdashline
\begin{array}{cccccc}
0 & 0 & 0 & \ldots\ldots\ldots\ldots & 0 & 0 \\
0 & 0 & 0 & \ldots\ldots\ldots\ldots & 0 & 0 \\
 & & & & & \\
0 & 0 & 0 & \ldots\ldots\ldots\ldots & 0 & 0
\end{array}
&
\begin{array}{cccc}
0 & 0 & \ldots\ldots\ldots & -b_2 \\
1 & 0 & \ldots\ldots\ldots & -b_{r-1} \\
0 & 1 & \ldots\ldots\ldots & -b_{r-2} \\
0 & 0 & \ldots\ldots\ldots\ 1 & -b_1
\end{array}
& &
\\ \hdashline
0 & 0 & & 0
\\ \hdashline
\begin{array}{cccccc}
0 & 0 & 0 & \ldots\ldots\ldots\ldots & 0 & 0 \\
 & & & & & \\
 & & & 0 & & \\
 & & & & & \\
 & & & & &
\end{array}
&
\begin{array}{ccc}
0 & 0 & \ldots\ldots\ldots \\
 & & \\
 & 0 & \\
 & & \\
 & &
\end{array}
&
\begin{array}{c} \\ \\ 0 \\ \\ \\ \end{array}
&
\begin{array}{ccccc}
0 & 0 & \ldots\ldots & & -g_v \\
1 & 0 & \ldots\ldots & & -g_{v-1} \\
0 & 0 & \ldots\ldots & 1 & -g_{v-2} \\
\ldots & \ldots & \ldots\ldots & & \\
0 & 0 & \ldots\ldots & 0 & -g
\end{array}
\end{array}
\right| \qquad (\alpha)
$$

where $P(\lambda)$ is the minimal polynomial of the whole space E_n. Since

$$P(M) = 0,$$

we get the important Cayley-Hamilton theorem:

(VIII-25) *Every square matrix* $[M]$ *satisfies its own characteristic equation.*

If the matrix $[M]$ is represented in an arbitrary basis $A_1, A_2, \dots, A_n$, we see that it is possible, by an appropriate change of basis, to give it the form (α). *This form is called the canonical form of the matrix* $[M]$.

If the field K over which the vector space E_n is defined contains all the roots of the various characteristic equations we must consider (this is true in particular if K is the field of complex numbers), we can always express the minimal polynomial of the subspaces $E_p, E_r, \dots, E_v$ as follows:

$$\begin{aligned} P(z) &= (z - z_1)^{p_1}(z - z_2)^{p_2} \dots (z - z_t)^{p_t} \\ R(z) &= (z - z_1)^{r_1}(z - z_2)^{r_2} \dots (z - z_t)^{r_t} \\ &\vdots \\ V(z) &= (z - z_1)^{v_1}(z - z_2)^{v_2} \dots (z - z_t)^{v_t}, \end{aligned}$$

where, of course,

$$\begin{aligned} &p_i > 0 \\ &p_i \geqslant r_i \geqslant \dots v_i \geqslant 0. \end{aligned}$$

The first condition expresses the fact that we have only taken the roots of $P(\lambda)$ into account, the second that the various polynomials $K(z)$ are divisors of the polynomials of greater rank and are divisible by the polynomials of smaller rank.

On the other hand, since the characteristic polynomial of $[M]$ is the product of the minimal polynomials $P(\lambda), R(\lambda), \ldots, V(\lambda)$, it follows that the quantities $\lambda_1, \lambda_2, \ldots, \lambda_t$ are the distinct roots of the characteristic equation.

On the other hand, the subspaces $E_p, E_r, \ldots,$ are cyclic and we have seen earlier that they can always be decomposed into direct sums of cyclic subspaces whose minimal polynomials are relatively prime, each minimal polynomial $P(\lambda), \ldots$ being the product of the corresponding irreducible minimal polynomials. Since the decomposition of the polynomial $P(\lambda)$ is unique, E_p decomposes into the direct sum of t cyclic subspaces with respectively $(\lambda - \lambda_1)^{\alpha_1}, (\lambda - \lambda_2)^{\alpha_2}, \ldots, (\lambda - \lambda_2)^{\alpha_t}$ as their minimal polynomials; and we have an analogous result for $E_r, \ldots, E_v$.

Since the subspaces defined in this way are invariant, if we consider the basis in the space E_n composed of k groups of vectors, where each group consists of a basis of one of the subspaces E_{k_j}, the matrix $[M]$ is written as follows:

$$M = \begin{bmatrix} [M_{p_1}] & & & & & \\ & \ddots & & & & \\ & & [M_{p_t}] & & & \\ & & & \ddots & & \\ & & & & [M_{r_1}] & \\ & & & & & \ddots \\ & & & & & & [M_{r_t}] \\ & & & & & & & \ddots \\ & & & & & & & & [M_{V_1}] \\ & & & & & & & & & \ddots \\ & & & & & & & & & & [M_{V_t}] \end{bmatrix}.$$

Consider one of the "partial" matrices $[M_{k_j}]$; it corresponds to the vector subspace E_{k_j} whose minimal polynomial is $(\lambda - z_j)^{k_j}$. There is a vector A in this subspace which admits $(\lambda - z_j)^{k_j}$ as its minimal polynomial. Consider then the following sequence of vectors:

$$A_1 = A, \qquad A_2 = (M - z_j I)\,A = (M - z_j I)\,A_1\,,$$

$$A_3 = (M - z_j I)^2 A = (M - z_j I)\,A_2, \ldots, A_{kj} = (M - z_j I)^{k_j - 1} A$$

$$= (M - z_j I)\,A_{k_{j-1}}.$$

The vectors $A_1, A_2, \ldots, A_{k_j}$ are linearly independent, since if they were not it would be possible to find a singular polynomial for A with degree smaller than k_j , which contradicts the fact that the minimal polynomial of A has degree k_j. These vectors $A_1, A_2, \ldots, A_{k_j}$ therefore form a basis of E_{k_j} and we have:

$$\begin{aligned}
[M]\,A_1 &= z_j\,A_1 + A_2 \\
[M]\,A_2 &= z_j\,A_2 + A_3 \\
&\vdots \\
[M]\,A_{k_j} &= z_j\,A_{k_j}.
\end{aligned}$$

Consequently, if we take the vectors $A_1, A_2, \ldots,$ as a basis of E_{k_j} the "partial" matrices $[M_{k_j}]$ are of the following form in this basis:

$$[M_{k_j}] = \begin{vmatrix}
z_1 & 0 & 0 & 0 & 0 \\
1 & z_1 & 0 & 0 & 0 \\
0 & 1 & z_1 & 0 & 0 \\
\cdot & \cdot & \cdot & \cdot & \cdot \\
0 & 0 & 1 & z_1 & 0 \\
0 & 0 & 0 & 1 & z_1
\end{vmatrix}.$$

If we define in the same way a basis in each of the

cyclic subsets forming the space E_n and take the set of bases formed in this way as the basis of E_n, the matrix $[M]$ finally assumes the following form in this new basis:

$$[M] = \begin{bmatrix} z_1 & & & & & & & & & & & & & \\ 1 & \ddots & & & & & & & & & & & & \\ & \ddots & z_1 & & & & & & & & & & & \\ & & 1 & \ddots & & & & & & & & & & \\ & & & \ddots & z_1 & & & & & & & & & \\ & & & & 1 & \ddots & & & & & & & & \\ & & & & & \ddots & z_2 & & & & & & & \\ & & & & & & 1 & \ddots & & & & & & \\ & & & & & & & \ddots & z_k & & & & & \\ & & & & & & & & 1 & \ddots & & & & \\ & & & & & & & & & \ddots & z_k & & & \\ & & & & & & & & & & 1 & \ddots & & \\ & & & & & & & & & & & \ddots & z_t & \\ & & & & & & & & & & & & 1 & \end{bmatrix},$$

where $z_1, z_2, .., z_t$ are the distinct roots of the characteristic equation and each root appears a number of times equal to its multiplicity. *We say that the matrix* $[M]$ *has been reduced to its Jordan canonical form.*

In the case where the roots of the characteristic polynomial $|[M] - \lambda[I]|$ are all simple, we have:

$$|[M] - \lambda[I]| = \prod(\lambda - z_i).$$

But we also have

$$|[M] - \lambda[I]| = P(\lambda)\, R(\lambda)\, ..., V(\lambda)$$

since each of the polynomials in the right-hand side must divide the polynomials of lower rank; whence the following conditions for the exponents of the binomials forming the $T(\lambda)$:

$$p_i = 1\,, \qquad\qquad i = 1, 2, ..., t$$
$$p_i + r_i + \cdots + v_i = 1\,, \qquad i = 1, 2, ..., t$$

which implies the following results:

$$r_i = 0,\ ...,\ v_i = 0, \qquad i = 1, 2, ..., t$$
$$\text{and} \quad t = n\,.$$

and the characteristic equation reduces to the following formula:

$$|\, M - \lambda I \,| = P(\lambda) = \prod(\lambda - z_i)\,.$$

Each one of the cyclic subspaces E_{k_j} considered above will therefore be of dimension 1 and we can take any vector A in it as a basis vector (moreover, in a space E_{k_j} all the vectors are identical up to length, since E_{k_j} is one-dimensional). Then

$$|\, [M] - \lambda[I] \,|\, A = \theta\,,$$

and therefore

$$[M]\, A = \lambda A\,.$$

The vector A is therefore one of the eigenvectors of the matrix $[M]$ and, consequently, taking the n eigenvectors of the matrix $[M]$ as a basis of E_n yields the well-known diagonal matrix already studied in Chapter VII.

CHAPTER IX

APPLICATION OF LINEAR ALGEBRA TO THE STUDY OF QUADRATIC FORMS

At the beginning of Chapter VI we introduced the concept of a bilinear function or form associated with every pair of vectors X, Y in the space E_n, and we introduced the scalar product of two vectors in terms of this concept.

We now take up this subject again and assume, as we have already done several times, that the space E_n is referred to an orthonormal basis.

Consider thus the following bilinear form associated with the vectors X and Y of E_n:

$$f(X, Y) = X'[A]\,Y = Y'[A']\,X.$$

The matrix $[A]$ is not arbitrary, for if we take $Y = X$ we find

$$f(X, X) = X'[A]\,X = X'[A]'\,X,$$

which implies

$$[A] = [A]'.$$

Indeed, if we expand these two expressions $X'[A]X$ and $X'[A]'X$, we find:

$$\sum_{i,j} a_i^j\, x_i\, x_j = \sum_{i,j} a_j^i\, x_i\, x_j\,.$$

(IX-1) *The matrix* $[A]$ *of the bilinear form is a symmetric matrix.*

In the rest of this chapter $[G]$ will denote symmetric matrices defining bilinear forms in matrix form.

(IX-2) *The quantity $X'[G]X$ introduced above, which is also denoted $Q(X, [G]X)$, is said to be the quadratic form associated with the bilinear function with matrix $[G]$. The rank of the matrix $[G]$ is said to be the rank of the quadratic form.*

We now establish some properties of quadratic forms which will be of great value for applications to mathematical statistics in general and to the analysis of linear models in particular.

(IX-3) *The sum of p quadratic forms $Q_1, Q_2, \dots, Q_p$ with rank $r_1, r_2, \dots, r_p$, respectively, is a quadratic form Q whose rank is at most equal to the sum of the ranks*

$$r_1 + r_2 + \cdots + r_p .$$

We have indeed

$$Q_i = X'[G_i]X ,$$

so that we can write

$$Q_1 + Q_2 + \cdots + Q_p = X'[G_1]X + X'[G_2]X + \cdots + X'[G_p]X$$

or

$$X'[G]X = X'([G_1] + [G_2] + \cdots + [G_p])X ,$$

where

$$[G] = [G_1] + [G_2] + \cdots + [G_p] .$$

It follows that the sum $Q_1 + Q_2 + \cdots + Q_p$ is indeed a quadratic form and its rank is that of the matrix $[G_1] + [G_2] + \cdots + [G_p]$. We know that the rank of this matrix is at most equal to the sum of the ranks of the matrices

$$[G_1], [G_2], \ldots, [G_p].$$

and this rank cannot be greater than n.

(IX-4) *A quadratic form which can be reduced to a sum of squares is called a fundamental quadratic form.*

We have seen earlier that a quadratic form can be written as follows:

$$X'[G]X = \sum_{i,j} g_i^j x_i x_j .$$

For a fundamental quadratic form we have

$$X'[G]X = \sum \lambda_i x_i^2 ,$$

which implies the following conditions for the matrix $[G]$:

$$\begin{aligned} g_i^j &= 0 \quad \text{for} \quad i \neq j \\ g_i^i &= \lambda_i \quad \text{for} \quad i = 1, 2, \ldots, n . \end{aligned}$$

In other words:

(IX-5) *The matrix of a fundamental quadratic form is a diagonal matrix and the coefficient of the square x_i^2 is simply the eigenvalue λ_i of this matrix.*

Now, we have shown in Chapter VII that, given a symmetric matrix $[G]$, there is a change of basis defined by an orthogonal matrix which transforms the matrix $[G]$ into a diagonal matrix whose diagonal consists of the simple or multiple eigenvalues of the original matrix $[G]$.

Two cases can occur:

a) The matrix $[G]$ *is non-singular.*

Then there exists an orthogonal change of basis defined by the matrix $[\Lambda]$ such that

$$[\Lambda]^{-1}[G][\Lambda] = \begin{vmatrix} \lambda_1 & & & 0 \\ & \lambda_2 & & \\ & & \ddots & \\ 0 & & & \lambda_n \end{vmatrix} = [G_\Lambda]$$

and if $x_{\Lambda i}$ denotes the components of the vector X with respect to the new basis, we find

$$X'[G]X = X'_\Lambda[G_\Lambda]X_\Lambda = \sum_i^n \lambda_i x_{\Lambda i}^2 .$$

b) The matrix $[G]$ *is of rank* r.

In this case, the change of basis with matrix $[\Lambda]$ transforms the matrix $[G]$ as follows:

$$[\Lambda]^{-1}[G][\Lambda] = \begin{bmatrix} \lambda_1 & & & & 0 \\ & \lambda_2 & & & \\ & & \ddots & & \\ & & & \lambda_r & 0 \\ 0 & & & & 0 \end{bmatrix} = [G_\Lambda]$$

and we get the following result:

$$X'[G]X = X'_\Lambda[G_\Lambda]X_\Lambda = \sum_{i=1}^r \lambda_i x_{\Lambda i}^2 .$$

These two results combine to form the following theorem:

(IX-6) *We can always find an orthogonal change of basis in the vector space* $\mathbf{E}_n$ *which transforms a quadratic form* $X'[G]X$ *into a sum of* n *squares if the quadratic form is non-singular, and into a sum of* r *squares if the quadratic form is of rank* r.

Since there are several changes of basis allowing us to transform a given quadratic form into a sum of squares, we now examine how such a change of basis affects the transformation. In this connection we now state Sylvester's Theorem.

(IX-7) *In the transformation of the quadratic form* $X'[G]X$ *into a sum of squares, the number of squares and the number of squares with positive coefficients are independent of the transformation.*

This result is obvious when we think of the matrix interpretation of these coefficients. These are in fact the roots of the characteristic equation of the matrix $[G]$, which remains invariant under a change of basis. The various possible transformations of the quadratic form $X'[G]X$ correspond to the various permutations of the roots k_i. But the number of roots, as well as their sign and the number of zero roots, is invariant.

We now introduce a special class of quadratic forms which are used very often in mathematical statistics.

POSITIVE DEFINITE QUADRATIC FORMS

(IX-8) *The quadratic form* $Q(X, [G]X)$ *is said to be positive definite if, for any* X *not identically zero,*

$$Q(X, [G]\,X) = \sum_{i,k} g_i^k\, x_i\, x_k > 0\,.$$

The quadratic form is said to be positive semi-definite if the sign > is replaced by ≥:

$$Q(X, [G]X) \geqslant 0 .$$

(IX-9) *A quadratic form is positive definite if and only if the eigenvalues of the matrix* $[G]$ *are all positive.*

The condition is necessary.

If the quadratic form Q is positive definite, we have, for any X not identically zero:

$$Q(X, [G]X) > 0 .$$

Now

$$Q(X, [G]X) = X'[G]X = X'_\Lambda G_\Lambda X_\Lambda$$

where

$$[G_\Lambda] = [\Lambda^{-1}][G][\Lambda] ,$$

that is

$$Q(X, [G]X) = k_1 x_{\Lambda 1}^2 + k_2 x_{\Lambda 2}^2 + \cdots + k_n x_{\Lambda n}^2$$

and we must have

$$k_1 x_{\Lambda 1}^2 + k_2 x_{\Lambda 2}^2 + \cdots + k_n x_{\Lambda n}^2 > 0 ,$$

for any $x_{\Lambda 1}, x_{\Lambda 2}, \ldots, x_{\Lambda n}$ not all identically zero. And this condition requires that all the k_i be positive. All the eigenvalues of $[G]$ are therefore positive and it follows in particular that $[G]$ is non-singular.

The condition is sufficient.

If all the eigenvalues of the matrix $[G]$ are positive, we have

$$Q(X, [G]X) = k_1 x_1^2 + k_2 x_2^2 + \cdots + k_n x_n^2 .$$

And this quantity is necessarily positive.

It follows from this condition that the matrix $[G]$ is non-singular and that its determinant is positive. We have indeed:

$$|G| = |\Lambda^{-1} G_\Lambda \Lambda| = |\Lambda^{-1}| |G_\Lambda| \lambda = |G_\Lambda| .$$

Now

$$|G_\Lambda| = k_1 k_2 \ldots k_n ,$$

hence

$$|G_\Lambda| = k_1 k_2 \ldots k_n > 0 .$$

Positive semidefinite forms have an analogous property:

(IX-10) *A quadratic form is positive semidefinite if and only if the matrix $[G]$ is singular and the non-zero roots of its characteristic equation are positive.*

This can be proved in exactly the same way as the previous case.

The quadratic form is written as follows

$$Q = k_1 x_{\Lambda 1}^2 + k_2 x_{\Lambda 2}^2 + \cdots + k_n x_{\Lambda n}^2$$

and this sum of squares is equal to zero without all the $x_{\Lambda i}$ being identically zero if and only if at least one of the roots k_i is zero; consequently, the matrix $[G_\Lambda]$ is singular and since it has the same rank as the matrix $[G]$, the latter is also singular.

We then see in the same way as previously that the non-zero eigenvalues are positive.

(IX-11) *If a variable x_i does not occur to the power 2 in a positive definite or semidefinite quadratic form $Q(X, [G]X)$, it does not occur to the power 1, that is, it does not occur in Q.*

In other words, if Q does not contain x_i^2, it contains no rectangular terms in x_i.

Since the form Q is positive definite or semidefinite, we always have

$$Q \geqslant 0$$

for any x_i.

Suppose now that a rectangular term in x_i is not equal to zero, that is, there is a coefficient a_i^k not equal to zero. Then let all the variables other than x_i and x_k be equal to zero; since x_i^2 does not occur in Q, we get

$$Q = a_i^k x_i x_k + a_k^k x_k^2$$

Now, a_k^k cannot be zero, for if it were we would have

$$Q = a_i^k x_i x_k$$

and we would only have to take x_i and x_k with the same sign if a_i^k is negative, or with opposite sign if a_i^k is positive, to get

$$Q < 0.$$

which contradicts the hypothesis that Q is positive semidefinite.

We now take a positive value for x_k and write

$$Q = x_k(a_i^k x_i + a_k^k x_k)$$

and it will be sufficient to take

$$x_i < \frac{-a_k^k}{a_i^k}$$

to get once more

$$Q < 0 .$$

Therefore, we must have

$$a_i^k = 0 .$$

N.B. The properties we have just given for positive definite and semidefinite quadratic forms can readily be carried over to negative definite and semidefinite forms which satisfy respectively the following inequalities:

$Q < 0$, negative definite quadratic form;

$Q \leq 0$, negative semidefinite quadratic form;

The theorem we are now going to state plays a fundamental role in the analysis of variance.

Cochran's Theorem

(IX-12) *If the quadratic form Q of rank r is written as the sum of k positive definite quadratic forms $Q_1, Q_2, \ldots, Q_k$ with ranks $r_1, r_2, \ldots, r_k$, respectively, where*

$$\sum_i r_i = n ,$$

and if the matrix $[G]$ associated with Q is non-singular and has all its eigenvalues equal to 1, there is an orthogonal change of basis such that the quadratic forms $Q_1, Q_2, \ldots, Q_k$ are replaced by sums of squares in such a way that each variable occurs in only one Q_i.

Suppose we have the following relation:

$$Q = Q_1 + Q_2 + \cdots + Q_k$$

where

$$Q = \sum_i \sum_k g_i^k x_i x_k$$

$$Q_1 = \sum_i \sum_k {}_1g_i^k x_i x_k$$

where the form Q is of rank n - that is, its matrix $[G]$ is non-singular - and the form Q_i is of rank r_i.

We show that the Q_i can be transformed as follows:

$$Q_1 = \sum_1^{r_1} x_{Ai}^2$$

$$Q_2 = \sum_{r_1+1}^{r_1+r_2} x_{Ai}^2$$

$$Q_p = \sum_{r_1+r_2+\cdots+r_{p-1}+1}^{r_1+r_2+\cdots+r_p} x_{Ai}^2$$

$$Q_k = \sum_{n-r_k+1}^{n} x_{Ai}^2 .$$

The proof is performed by induction, showing that the result is true for k if it is true for k - 1.

We first perform a change of basis such that the eigenvectors of $[G]$ are the new basis vectors; since all the eigenvalues are equal to 1, we have

$$Q = \sum \sum g_i^k x_i x_k = \sum_i x_{Ai}^2 .$$

Under this transformation, each one of the quadratic forms Q_i is transformed into another form of same rank r_i, since

$$Q_1 = X'[G_i] X = X_A'[A]^{-1} [G] [A] X_A ,$$

and the matrix $[\Lambda]^{-1}[G][\Lambda]$ has the same rank as the matrix $[G]$.

We can therefore write, with the new basis:

$$\sum_{i=1}^{i=n} x_{\Lambda i}^2 = Q_1' + Q_2' + \cdots + Q_k'$$

where the primes in the right-hand side indicate that the quadratic forms are referred to the new basis.

If $k = 1$, the theorem is obvious since

$$Q_1' = \sum_{i=1}^{n} x_{\Lambda i}^2 .$$

We now show that if the theorem is true for $k - 1$ it is true for k.

We perform a new change of basis, taking as basis vectors the r_1 eigenvectors of the matrix associated with Q_1' completed with $n - r_1$ vectors so as to form an orthonormal basis; Q_1' then becomes Q_1'', where

$$Q_1'' = \sum_{1}^{r_i} \mu_i x_{mi}^2 .$$

Since the matrix $[M]$ of the change of basis is orthogonal, the sum

$$\sum_{i=1}^{n} x_{\Lambda i}^2$$

is transformed into

$$\sum_{i=1}^{n} x_{Mi}^2$$

and each one of the quadratic forms Q_i' is transformed into a quadratic form Q_i'' of same rank r_i. We can therefore write

$$\sum_{1}^{n} x_{Mi}^2 = \sum_{1}^{r_1} \mu_i x_{Mi}^2 + Q_2'' + \cdots + Q_k''$$

or

$$\sum_{1}^{r_1} x_{Mi}^2(1 - \mu_i) + \sum_{r_1+1}^{n} x_{Mi}^2 = Q_2'' + \cdots + Q_k''.$$

Now, in view of Theorem (IX-3), the rank of the sum $Q_2'' + \cdots + Q_k''$ in the right-hand side is at most equal to $n - r_1$.

But, in the left-hand side, the sum $\sum_{r_1+1}^{n} x_{Mi}^2$ is of rank $n - r_1$. The sum $\sum_{i}^{r_1} x_{Mi}^2(1 - \mu_i)$ must therefore be of rank 0, that is,

$$\mu_1 = \mu_2 = \cdots = \mu_{r_1} = 1.$$

In other words, Q_1 can be written as follows:

$$Q_1'' = \sum_{1}^{r_1} x_{Mi}^2.$$

And we can also write

$$\sum_{r_1+1}^{n} x_{Mi}^2 = Q_2'' + \cdots + Q_k''.$$

None of the forms Q_i'' contains the variables

$$x_{M1}, x_{M2}, \ldots, x_{Mr_1}.$$

For suppose, say, that Q_2'' contains a term x_{Mi} with $i < r_1$. This term will occur to the power 2 in Q_2'' since Q_2'' is positive semidefinite (Theorem IX-11), with a positive coefficient. It can only occur in the other forms Q_i'' with a non-negative coefficient. The above equality would then be impossible since x_{Mi} would occur in the right-hand side and not in the left-hand side.

We have therefore decomposed the sum $Q_1' + Q_2' + \cdots + Q_k'$ into the expression

$$\sum_{1}^{r_1} x_{Mi}^2 + Q_2'' + \cdots + Q_k''$$

where none of the variables occuring in the first form occurs also in the forms $Q_2'', \ldots, Q_k''$.

Now, the theorem is true for the k - 1 forms $Q_2'' + \cdots + Q_k''$. There is therefore an orthogonal transformation with matrix $[N_1]$ such that

$$Q_2'' + \cdots + Q_k'' = \sum_{r_1+1}^{r_1+r_2} x_{Mi}^2 + \cdots + \sum_{n-r_k+1}^{n} x_{Ni}^2 .$$

Since this transformation is orthogonal, it does not affect the first sum $\sum_1^{r_1} x_{Mi}^2$, nor the sum in the left-hand side, $\sum_1^n x_{Mi}^2$.

The last change of basis transformed the variables $x_{Mr_1+1}, \ldots, x_{Mn}$ into the variables $x_{Nr_1+1}, \ldots, x_{Nn}$. We now complete this change of basis by setting:

$$x_{M1} = x_{N1}, x_{M2} = x_{N2} \ldots, x_{Mr_1} = x_{Nr1} .$$

This yields a change of basis defined by the following matrix:

$$[N] = \begin{array}{r} r_1\left\{\vphantom{\begin{matrix}1\\1\\1\end{matrix}}\right. \\ \vphantom{N_1} \end{array} \left[\begin{array}{c|c} \overbrace{\begin{matrix}1 & & 0\\ & 1 & \\ 0 & & 1\end{matrix}}^{r_1} & 0 \\ \hline 0 & N_1 \end{array}\right].$$

It is readily shown that the matrix $[N]$ is orthogonal. Consider its transpose

$$[N'] = \begin{array}{r} r_1\left\{\vphantom{\begin{matrix}1\\1\\1\end{matrix}}\right. \\ \vphantom{N_1'} \end{array} \left[\begin{array}{c|c} \overbrace{\begin{matrix}1 & & 0\\ & 1 & \\ 0 & & 1\end{matrix}}^{r_1} & 0 \\ \hline 0 & N_1' \end{array}\right].$$

Hence $[N]'[N] = [I]$. For the rows 1 to r_1 have all their elements equal to zero except those on the main diagonal which are equal to 1. We now look for the elements of the row $r_1 + 1$. We have

$$P_{R_1+1} = \sum_{i=1}^{n} (n_{r_1+1}^i)' n_i^k = \sum_{r_1+1}^{n} (n_{r_1+1}^i)' n_i^k = \begin{cases} 1 \text{ if } k = r_1 + 1, \\ 0 \text{ if } k \neq r_1 + 1, \end{cases}$$

since the matrix $[N_1]$ is orthogonal.

Therefore the matrix $[N]'[N]$, where all elements are zero except those on the main diagonal, which are equal to 1, is the identity matrix and $[N]$ is indeed an orthogonal matrix.

We have thus gone from the system

$$Q = Q_1 + \cdots + Q_n$$

to the system

$$\sum_1^n x_{Ni}^2 = \sum_1^{r_1} x_{Ni}^2 + \sum_{r_1+1}^{r_1+r_2} x_{Ni}^2 + \cdots + \sum_{n-r_k+1}^{n} x_{Ni}^2$$

after three orthogonal changes of basis

$$X \to [\Lambda] X = X_\Lambda , \qquad X_\Lambda \to [M] X_\Lambda = X_M , \qquad X_M \to [N] X_M = X_N ,$$

that is,

$$X_N = [N] X_M = [N][M] X_\Lambda = [N][M][\Lambda] X .$$

The triple change of basis is equivalent to a unique change of basis defined by the matrix $[N][M][\Lambda]$, which is orthogonal since the matrices $[N]$, $[M]$ and $[\Lambda]$ are all orthogonal.

We have thus proved Cochran's Theorem.

THE MATRIX OF AN ANALYTICALLY DEFINED QUADRATIC FORM

We saw at the beginning of this chapter that to every symmetric matrix $[G]$ with n rows and n columns there corresponds a quadratic form defined as follows:

$$Q(X, [G]\, X) = X'[G]\, X\,.$$

The quadratic form is often given in the usual algebraic form

$$Q(x_1, x_2, ..., x_n) = \sum_{i,j} a_i^k\, x_i\, x_k\,,$$

and it is then desirable to be able to express it in the matrix form which affords all the analyses and simplifications we have just examined.

We immediately see that

$$a_i^k = \frac{\partial Q}{\partial x_i\, \partial x_k}\,,$$

which shows once more that the matrix $[G]$ is symmetric. Indeed:

$$a_k^i = \frac{\partial Q}{\partial x_k\, \partial x_i} = \frac{\partial Q}{\partial x_i\, \partial x_k} = a_i^k\,.$$

We can therefore state the following result:

(IX-13) *Every quadratic form $Q(x_1, x_2, ..., x_n)$ can be expressed in matrix form $X'[G]X$, where the element in the ith row and the kth column of the matrix $[G]$ is the second-order partial derivative of Q with respect to x_i and x_k.*

CHAPTER X

APPLICATION OF LINEAR ALGEBRA TO STATISTICS AND TO THE CALCULUS OF PROBABILITIES

This chapter covers several applications of linear algebra to statistics and to the calculus of probabilities. We have chosen them to be rather varied, in order to show the wide range of application of the methods of linear algebra and to emphasize their usefulness in understanding random processes as soon as the somewhat tedious initial calculations have been taken care of. Moreover, if the calculations performed by means of linear algebra are not always simple, these same computations would be practically impossible by the classical methods.

This chapter is divided into three sections which correspond to the three applications we have chosen.

I. MULTILINEAR REGRESSIONS

The following problem often arises in physics, technology and economics:

A phenomenon Y depends on several "explanatory" factors X_1, X_2, ..., and we would like to find a representative model allowing us to express Y in terms of the X_i, taking into consideration the fact that even if the dependence between Y and the X_i is functional, the representation observed is not, in view of the "errors" and "chances" which are unavoidable in the physical world.

Many types of models of varying intricacy have

been set up to represent this situation. Without dwelling on the very interesting general considerations, which would lie beyond the scope of this book, we shall give a brief description of a model which is very commonly used.

In the rest of this chapter we assume that the reader is familiar with the concepts of random variable, probability law, distribution function, mathematical expectation, etc. The model is then as follows:

Suppose that the phenomenon Y depends in a random fashion on the explanatory factors $X_1, X_2, \ldots, X_{n-1}$. We are looking for the "best" representation of Y in terms of $X_1, X_2, \ldots, X_{n-1}$. At this stage we must define the model in two ways:

a) by specifying the structure of the representation;

b) by fixing the criterion for the choice of the model.

STRUCTURE

We only consider the case where the model is linear. We then make a slight change of notation, calling the phenomenon X_1 and the explanatory factors $X_2, X_3, \ldots, X_n$.

We are therefore looking for a representation of the following form:

$$X_1^* = \alpha_1^2 X_2 + \alpha_1^3 X_3 + \cdots + \alpha_1^n X_n$$

where the α_1^j are chosen so that X_1^* is as "close" as possible to X_1. This proximity condition is in turn defined by the chosen criterion.

CRITERION OF CHOICE

The quantity X_1^* is a "prediction" of the phenomenon X_1 in terms of the factors X_2, X_3, ..., X_n; we can also say that it is an estimate of X_1. We impose upon this estimate the following condition:

$$E(X_1 - X_1^*)^2 = \text{minimum},$$

where the symbol $E(X)$ denotes the mathematical expectation of the random variable X.

This condition allows us to evaluate the coefficients α_j^i and therefore to define the model completely.

We assume in the sequel that the quantities X_1, X_2, ..., X_n obey a probability law with n variables defined by its distribution function:

$$F(x_1, x_2, ..., x_n) = \Pr[X_1 \leqslant x_1,\ X_2 \leqslant x_2,\ ..., X_n \leqslant x_n].$$

The only hypothesis we make is that the first and second moments of this probability law exist. We suppose without loss of generality that the first moments are zero. This simply amounts to replacing the variables X_i by their centred values.

Then

$$E(X_i) = \int_{E_n} x_i \, \mathrm{d}F(x_1, x_2, ..., x_n) = 0.$$

$$E(X_i^2) = \int_{E_n} x_i^2 \, \mathrm{d}F(x_1, x_2, ..., x_n) = v_i^i = \sigma_i^2,$$

$$E(X_i X_k) = \int_{E_n} x_i x_k \, \mathrm{d}F(x_1, x_2, ..., x_n) = v_i^k.$$

We know that v_i^i is the *variance* of X_i and v_i^k the *covariance* of X_i and X_k We also set

$$\frac{v_i^k}{\sqrt{v_i^i v_k^k}} = \rho_i^k$$

and ρ_i^k is the coefficient of total linear correlation of the variables X_i and X_k.

We now associate with the probability law under consideration an $n \times n$ matrix which will play a fundamental role in the rest of the exposition. It is the covariance matrix $[V]$ of the law $(X_1, X_2, \ldots, X_n)$:

(X-1)
$$[V] = \begin{bmatrix} v_1^1 & v_1^2 & \ldots & v_1^n \\ v_2^1 & v_2^2 & \ldots & v_2^n \\ \cdot & \cdot & \cdot & \cdot \\ v_n^1 & v_n^2 & \ldots & v_n^n \end{bmatrix}.$$

Obviously
$$v_i^k = v_k^i,$$

which implies that the matrix $[V]$ is symmetric.

(X-2) *Moreover, the matrix $[V]$ is the matrix of a positive semidefinite quadratic form.*

This is readily seen by considering the following quantity, in which $t_1, t_2, \ldots, t_n$ are parameters which can assume arbitrary values:

$$E\left[\left(\sum t_i X_i\right)^2\right] = E\left[\sum_{i,k} t_i t_k X_i X_k\right] = \sum_{i,k} t_i t_k E(X_i X_k) = \sum_{i,k} t_i t_k v_i^k.$$

The matrix $[V]$ is the matrix of the quadratic form $E[(\Sigma t_i X_i)^2]$. Now, the mathematical expectation of a square is always positive or zero, and therefore

$$\sum_{i,k} t_i t_k v_i^k \geqslant 0,$$

which proves the property.

Furthermore, we define the correlation matrix as follows:

$$(X-3) \qquad [R] = \begin{bmatrix} 1 & \rho_1^2 & \dots & \rho_1^n \\ \rho_2^1 & 1 & \dots & \rho_2^n \\ \cdot & \cdot & \cdot & \cdot \\ \rho_n^1 & \rho_n^2 & \dots & 1 \end{bmatrix}.$$

Starting from the definition of the correlation coefficient

$$\rho_i^k = \frac{v_i^k}{\sqrt{v_i^i v_k^k}} = \frac{v_i^k}{\sigma_i \sigma_k}$$

we immediately deduce the following relation between the matrices $[V]$ and $[R]$:

$$[V] = [S][R][S],$$

where $[S]$ is the diagonal matrix

$$(X-4) \qquad [S] = \begin{bmatrix} \sigma_1 & 0 & \dots & 0 \\ 0 & \sigma_2 & \dots & \\ \cdot & \cdot & \cdot & \cdot \\ 0 & 0 & \dots & \sigma_n \end{bmatrix}.$$

Indeed, the matrix $[R]$ $[S]$ can be deduced from $[R]$ by multiplying the first column by σ_1, the second by σ_2, ..., the nth by σ_n. $[S]$ $[R]$ $[S]$ can be deduced from $[R]$ $[S]$ by multiplying the first row by σ_1, the second by σ_2, ..., the nth by σ_n; it follows that

$$(X-5) \qquad [S][R][S] = \begin{bmatrix} \sigma_1^2 \rho_1^1 & \sigma_1 \sigma_2^2 \rho_1^2 & \dots & \sigma_1 \sigma_n \rho_1^n \\ \sigma_1 \sigma_2 \rho_2^1 & \sigma_2^2 \rho_2^2 & & \sigma_2 \sigma_n \rho_2^n \\ \cdot & \cdot & \cdot & \cdot \\ \sigma_1 \sigma_n \rho_n^1 & \sigma_2 \sigma_n^1 \rho_n^2 & \dots & \sigma_n^2 \rho_n^2 \end{bmatrix},$$

and the covariances are the general elements of the matrix [S] [R] [S]. We now formulate the problem in its final form.

The study of the linear regression of a random variable X_1 *in terms of* $n-1$ *other variables* X_2, X_3, ..., X_n, *given the probability law of the set* X_1, X_2, ..., X_n *or more simply the first and second moments of this law, consists in finding a linear combination* X_1^* *of the variables* X_2, X_3, ..., X_n *such that the mathematical expectation* $E(X_1 - X_1^*)^2$ *of the square of the difference between the variable and its estimate is minimal.*

DETERMINATION OF THE COEFFICIENTS IN X_1^*

Consider the expression $T = E[(X_1 - X_1^0)^2]$,

where $$X_1^0 = \alpha_1^2 X_2 + \alpha_1^3 X_3 + \cdots + \alpha_1^n X_n .$$

We have $$T = \int_{E_n} (x_1 - \alpha_1^2 x_2 - \alpha_1^3 x_3 - \cdots - \alpha_1^n x_n)^2 \, dF .$$

A necessary condition for this quantity to be an exetremum is that its partial derivatives with respect to the coefficients α_i^k vanish:

$$\frac{\partial T}{\partial \alpha_1^k} = 0 \text{ for } k = 2, 3, \ldots, n .$$

which implies here that

$$\frac{\partial T}{\partial \alpha_1^k} = \int_{E_n} - 2\, x_k(x_1 - \alpha_1^2 x_2 - \alpha_1^3 x_3 - \cdots - \alpha_1^n x_n) \, dF = 0$$
$$\text{for } k = 2, 3, \ldots, n .$$

Performing the integration yields

$$-2(v_k^1 - \alpha_1^2 v_k^2 - \alpha_1^3 v_k^3 - \cdots - \alpha_1^n v_k^n) = 0 \quad \text{for } k = 2, 3, \ldots, n,$$

that is, rearranging the terms and writing all the corresponding equations:

$$(\alpha) \begin{cases} v_2^2 \alpha_1^2 + v_2^3 \alpha_1^3 + \cdots + v_2^n \alpha_1^n = v_2^1 \\ v_3^2 \alpha_1^2 + v_3^3 \alpha_1^3 + \cdots + v_3^n \alpha_1^n = v_3^1 \\ \vdots \\ v_n^2 \alpha_1^2 + v_n^3 \alpha_1^3 + \cdots + v_n^n \alpha_1^n = v_n^1 \end{cases}$$

The coefficients α_1^k are therefore determined by a system of $n - 1$ equations in $n - 1$ unknowns whose *matrix is the minor of the term v_1^1 in the matrix $[V]$ introduced earlier, on page 196. The determinant of this minor is denoted V_1^1.*

Two cases may arise.

First case. The quadratic form $E[(\sum t_i X_i)^2]$ is positive definite, that is, the probability law of the set of the variables $X_1, X_2, \ldots, X_n$ is non-degenerate. The determinant of the matrix $[V]$ and its principal minors are then positive, in particular V_1^1, and the system (α) is a Cramer system. We therefore write:

This column is in the position of the kth column.

$$\alpha_1^k = \frac{\begin{bmatrix} v_2^2 & v_2^3 & \dots & v_2^1 & v_2^n \\ v_3^2 & v_3^3 & \dots & v_3^1 & v_3^n \\ \cdot & \cdot & \cdot & \cdot & \cdot \\ v_n^1 & v_n^2 & \dots & v_n^1 & v_n^n \end{bmatrix}}{\begin{bmatrix} v_2^2 & v_2^3 & \dots & \dots & v_2^n \\ v_3^2 & v_3^3 & \dots & \dots & v_3^n \\ \cdot & \cdot & \cdot & \cdot & \cdot \\ v_n^2 & v_n^3 & \dots & \dots & v_n^n \end{bmatrix}} = \frac{D_1^k}{V_1^1}.$$

We now perform $k - 1$ permutations on the column $\begin{bmatrix} v_2^1 \\ v_3^1 \\ \vdots \\ v_n^1 \end{bmatrix}$ in the determinant D_1^k, in order to put it in first position. We get:

$$D_1^k = (-1)^{k-2} \begin{bmatrix} v_2^1 & v_2^2 & \dots & v_2^n \\ v_3^1 & v_3^2 & \dots & v_3^n \\ \cdot & \cdot & \cdot & \cdot \\ v_n^1 & v_n^2 & \dots & v_n^n \end{bmatrix} = (-1)^{k-2} D_1^{k'}.$$

We immediately see that this determinant D_1^k can be deduced from the determinant of $[V]$ by removing the first row and the kth column. It is thus simply the minor V_1^k, divided by $(-1)^{1+k}$, of the term v_1^k in the determinant of $[V]$, that is,

$$D_1^{k'} = \frac{V_1^k}{(-1)^{1+k}}.$$

Hence finally

$$D_1^k = \frac{(-1)^{k-2}}{(-1)^{k+1}} \quad V_1^k = -V_1^k.$$

We therefore find the following simple formula for the coefficients α_1^k:

(X-6)
$$\boxed{\alpha_1^k = -\frac{V_1^k}{V_1^1}}$$

and the estimate X_1^* is of form

(X-7)
$$\boxed{X_1^* = -\frac{1}{V_1^1}\sum_k V_1^k X_k}$$

We then find

$$\boxed{E(X_1^*) = 0}$$

(X-8)
$$\boxed{\operatorname{var} X_1^* = v_1^1 - \frac{|V|}{V_1^1}} .$$

This second formula will be proved later.

We must now verify that substitution of the linear combination X_1^* in the expression $E(X_1 - X_1^0)^2$ yields a minimum and not a maximum. We examine the sign of the expression

$$E(X_1 - X_1^0)^2 - E(X_1 - X_1^*)^2$$

in the neighbourhood of the point X_1^*. Expanding the function T introduced above we find

$$E(X_1 - X_1^0)^2 - E(X_1 - X_1^*)^2 = \sum_{i,j} \frac{\partial^2 T}{\partial \alpha_1^k \, \partial \alpha_1^j} (\alpha_1^{k\text{o}} - \alpha_1^{k*})(\alpha_1^{j\text{o}} - \alpha_1^{j*})$$

in the neighbourhood of the point X_1^*. Now

$$\frac{\partial^2 T'}{\partial \alpha_1^k \, \partial \alpha_1^j} = \int_{E_n} x_k \, x_j \, \mathrm{d}F = 2 \, v_j^k$$

and therefore

$$E[(X_1 - X_1^0)^2] - E[(X_1 - X_1^*)^2] = \sum_{i,j} 2 \, v_j^k (\alpha_1^{k0} - \alpha_1^{k*}) (\alpha_1^{j0} - \alpha_1^{j*}) .$$

This is a quadratic form Q whose matrix, up to a factor 2, is the matrix $[V]$ with its first row and its first column removed (i and j cannot be 1). Since the quadratic form $E[(\sum t_i X_i)^2]$ is positive definite, so is the form derived from it by removing the variable X_1. It follows that Q is always positive and the expression $E(X_1 - X_1^0)^2$ is indeed minimal for $X_1^0 = X_1^*$.

Second case. The quadratic form $E[(\sum t_i X_i)^2]$ is positive semidefinite, that is, the probability law of the set $X_1, X_2, \ldots, X_n$ is degenerate. The matrix $[V]$ is of rank r.

The results we have just obtained are still valid, apart from a few small changes.

We can indeed still consider the first row of the matrix $[V]$ as a linear combination of the $n - 1$ others, that is, in fact, of the r linearly independent columns. The matrix $[V_1^1]$ is also of rank r. The system (α) therefore reduces to r equations and is always possible since the right-hand sides consist of the first column of $[V]$ and therefore satisfy the compatibility conditions.

The regression then contains only r terms in X_i.

In the sequel, and for simplicity, we always assume that the matrix $[V]$ is non-singular, so that the associated quadratic form is positive definite. If this were not the case, it would be sufficient, in order to get back to this case, to extract from $[V]$ a non-zero minor of rank r containing the first column.

PROPERTIES OF THE ESTIMATE X_1^*

The estimate X_1^* has some interesting properties, connected with the concept of linear correlation coefficient introduced at the very beginning.

(X-9) *The linear correlation coefficient of the difference $Z_1 = X_1 - X_1^*$ with each of the variables $X_2, \ldots, X_n$ is zero.*

Note first that
$$E(Z_1) = E(X_1 - X_1^*) = 0\,.$$

It follows that the covariance of X_k with Z_1 can be written as:
$$E(X_k Z_1) = E\left[X_k\left(X_1 - \sum_{j-2}^{n} \alpha_1^j X_j\right)\right],$$

that is,
$$E(X_k Z_1) = v_k^1 - \sum_2^n \alpha_1^j v_k^1 = v_k^1 + \frac{1}{V_1^1}\sum_2^n V_1^j v_k^1$$

or
$$E(X_k Z_1) = \frac{1}{V_1^1}\sum_{j=1}^{n} V_1^j v_k^j \quad .$$

The sum in the right-hand side represents the expansion with respect to the kth row of a determinant identical to the determinant of $[V]$ but in which the kth row is identical to the first; it is therefore equal to zero and we have
$$E(X_k Z_1) = 0\,, \qquad k = 2, \ldots, n\,.$$

Since the covariance is zero, the same is obviously true of the linear correlation coefficient.

(X-10) *The estimate X_1^* is that linear combination of the variables $X_2, X_3, \ldots, X_n$ for which the linear correlation coefficient with X_2 is greatest.*

We first show that the expression $E(X_1^{0\,2}) - E(X_1 X_1^0)$ is always positive unless $X_1^0 = X_1^*$, when it is zero.

We have

$$E(X_1^{0\,2}) - E(X_1\,X_1^0) = \sum_{j,k} \alpha_1^j\,\alpha_1^k\,v_j^k - \sum_k \alpha_1^k\,v_1^k = R(X_1^0)\,.$$

As we have already done earlier, we expand $R(X_1^0)$ in the neighbourhood of the point X_1^*, which gives us

$$R^0 - R^* = \sum_j \frac{\partial R}{\partial \alpha_1^j}(\alpha_1^{j0} - \alpha_1^{j*}) + \sum_{j,k} \frac{\partial^2 R}{\partial \alpha_1^j\,\partial \alpha_1^k}(\alpha_1^{j0} - \alpha_1^{j*})(\alpha_1^{k0} - \alpha_1^{k*})\,.$$

Now, at the point X_1^*,

$$\frac{\partial R}{\partial \alpha_1^k} = \sum_j \alpha_1^j\,v_j^k - v_1^k \qquad \frac{\partial^2 R}{\partial \alpha_1^j\,\partial \alpha_1^k} = v_j^k\ .$$

Referring to the proof of the preceding theorem we immediately see that the first-order partial derivatives are equal to zero, which implies

$$R^0 - R^* = \sum_{j,k} v_j^k(\alpha_1^k - \alpha_1^{k*})(\alpha_1^j - \alpha_1^{j*})\,.$$

We are dealing with a positive definite form, that is, we have always

$$R^0 - R^* \geqslant 0\,, \quad \text{that is,} \quad R^0 \geqslant R^*\,.$$

If follows that R^0 is a minimum for $X_1^0 = X_1^*$, and we have

$$R^* = E\,[X_1^{*2}] - E\,[X_1^*\,X_1] = E\,[X_1^*(X_1^* - X_1)] = E\,[X_1^*\,Z_1] = 0$$

since Z_1 has zero correlation coefficient with each of the X_j of which X_1 is a linear combination.

We therefore always have

$$E(X_1^{02}) - E(X_1\,X_1^0) \geqslant 0$$

or

$$E(X_1\,X_1^0) \leqslant E(X_1^{02})\,.$$

Therefore, if we denote by ρ_{10} the linear correlation coefficient between X_1 and X_1^0, we get

$$\rho_{10} = \frac{E(X_1\,X_1^0)}{\sqrt{E(X_1^2)\,E(X_1^{02})}} \leqslant \frac{E(X_1^{02})}{\sqrt{E(X_1^0)\,E(X_1^{02})}}\ \sqrt{\frac{E(X_1^{02})}{E(X_1^2)}}\,.$$

It follows that ρ reaches its maximum value when $X_1^0 = X_1^*$, and this value is

$$\sqrt{\frac{E(X_1^{*2})}{E(X_1^2)}}\,.$$

We will now use these two properties to introduce the concepts of multiple correlation coefficient and partial correlation.

We first note that the roles of the variables X_1, X_2, ..., X_n are interchangeable and that the formulas allowing us to estimate X_1 in terms of

$X_2, \ldots, X_n$ can readily be carried over to estimate X_i in terms of $X_1, X_2, \ldots, X_{i-1}, X_{i+1}, \ldots,$ by simply replacing everywhere the index 1 by the index i. For the sake of generality, we now suppose that the variable X_i is estimated in terms of $X_1, X_2, \ldots, X_n$.

MULTIPLE CORRELATION COEFFICIENT

We have just seen that X_i^* is the linear combination of the variables $X_1, X_2, \ldots, X_n$ with greatest correlation with X_i.

It is therefore natural to say that

(X-11) *The ordinary correlation coefficient (the total correlation coefficient)* ρ_i *between* X_i *and* X_i^* *represents the correlation between* X_i *and the set of the variables* $X_1, X_2, \ldots, X_n$. *This total correlation coefficient between* X_i *and* X_i^* *is the multiple correlation coefficient between* X_i *on the one hand and the set of variables* $X_1, X_2, \ldots, X_n$ *on the other.*

We have thus
$$\rho_i = \frac{E(X_i X_i^*)}{\sqrt{E(X_1^2)\, E(X_1^{*2})}} .$$

Now we have seen earlier that
$$E[X_k X_i - X_i^*] = \frac{1}{V_i^i} \sum_j V_i^j v_k^j = 0 .$$

We also have
$$E\left[X_i(X_i - X_i^*)\right] = \frac{1}{V_i^i} \sum_j V_i^j v_i^j = \frac{|V|}{V_i^i},$$

that is,

$$E(X_i^2) - E(X_i X_i^*) = \frac{|V|}{V_i^i}.$$

It follows that

$$E(X_i X_i^*) = v_i^i - \frac{|V|}{V_i^i}.$$

On the other hand

$$E[X_i^*(X_i - X_i^*)] = 0$$

since X_i is a linear combination of the X_j and $X_i - X_i^*$ has xero correlation with each of the X_j. Hence

$$E(X_i X_i^*) = E(X_i^{*2})$$

We therefore immediately get

$$\rho_i = \frac{E(X_i X_i^*)}{\sqrt{E(X_i^2)\, E(X_i^2)}} = \frac{E(X_i^{*2})}{\sqrt{E(X_i^2)\, E(X_i^2)}} = \sqrt{\frac{E(X_i^{*2})}{E(X_1^2)}}$$

or

$$\rho_i = \sqrt{\frac{v_i^i - |V|/V_i^i}{v_i^i}} = \sqrt{1 - \frac{|V|}{v_i^i V_i^i}},$$

that is,

$$\rho_i = \sqrt{1 - \frac{|V|}{v_i^i V_i^i}}.$$

PARTIAL CORRELATION COEFFICIENT

In order to study the relationship between two variables X_i and X_k, we can use the ordinary correlation ρ_i^k, which we called earlier the total correlation coefficient. But what exactly is this coefficient ρ_i^k? We do not know if it indicates an actual relationship between X_i and X_k or if it simply expresses the fact that X_i and X_k each depend on the n - 2 other variables $X_1, X_2, \ldots, X_n$ and depend on one another through these n - 2 variables.

We have seen earlier that the differences Z_i are independent of the variables $X_1, X_2, \ldots, X_n$. It is therefore very natural to consider the ordinary correlation coefficient between Z_i and Z_k in order to measure the intrinsic dependence between X_i and X_k. In this case, obviously, the linear combinations X_i^* and X_k^* involve only the n - 2 variables different from X_i and X_k. This is emphasized by denoting these linear combinations $X_{i.k}^*$ and $X_{k.i}^*$.

(X-12) *The ordinary correlation coefficient between the differences $X_i - X_{i.k}^*$ and $X_k - X_{k.i}^*$, where the linear combinations $X_{i.k}^*$ and $X_{k.i}^*$ involve the n - 2 other variables $X_1, X_2, \ldots, X_n$, is called the partial correlation coefficient between X_i and X_k.*

We are now going to evaluate this partial correlation coefficient $\rho_{i.n-2}^k$.

Note first that the estimates X_i and X_k are slightly different from those we gave earlier, since they only involve n - 2 variables. We have

$$X_{i.k}^* = \sum_j \alpha_{i.k}^i X_j \ . \ X_{k.i}^* = \sum_j \alpha_{k.i}^i X_j$$

where

$$\alpha_{i.k}^i = -\frac{V_{i.kk}^j}{V_{i.kk}^i}$$

and
$$\alpha^j_{k.i} = -\frac{V^j_{k.ii}}{V^k_{k.ii}}.$$

The repeated symbol "kk" means that we no longer consider the minors of $|V|$ but the minors of the minor V^k_k. The system used to define the $\alpha^k_{i.k}$ is equivalent to the system considered above but does not contain the coefficient corresponding to X_k or the equation which represented the partial derivative of $E(X_i - X_i^*)^2$ with respect to α^k_i. It follows that we must replace V^j_i in the expression $\alpha^j_i = -V^j_i/V^i_i$ by an analogous determinant in which the kth row and the kth column have been removed: $V^j_{i.kk}$. The same is true for V^i_i. And the same remark is valid for the coefficients $\alpha_{k.i}$:

$$\rho^k_{i.n-2} = \frac{E(Z_{i.k}\,Z_{k.i})}{\sqrt{E(Z^2_{i.k})\,E(Z^2_{k.i})}}$$

$$Z_{i.k} = X_i - X^*_{i.k}, \qquad Z_{k.i} = X_k - X^*_{k.i}.$$

We have

$$E[X_i(X_i - X^*_{i.k})] = E\left[X_i^2 + \sum_j \frac{V^j_{i.kk}}{V^i_{i.kk}} X_i X_j\right] = \sum_j \frac{V^j_{i.kk}}{V^i_{i.kk}} V_i = \frac{V^k_k}{V^i_{i.kk}}.$$

Similarly

$$E[X_k(X_k - X^*_{k.i})] = V^i_i/V^k_{k.ii}.$$

But, on the other hand, from the theorem proved earlier, $X_i - X^*_{i.k}$ is independent of the X_j, and hence of $X^*_{i.k}$, and we have

$$E[(X_i - X^*_{i.k})\,X^*_{i.k}] = 0,$$

that is, on the one hand

$$E(X_{i.k}^{*2}) = E(X_i X_{i.k}^*),$$

and on the other hand

$$\begin{aligned} E(Z_{i.k}^2) &= E[(X_i - X_{i.k}^*)(X_i - X_{i.k}^*)] \\ &= E[X_i(X_i - X_{i.k}^*)] - E[X_{i.k} - X_{i.k}^*] \\ &= E[X_i(X_i - X_{i.k}^*)] - 0 = V_k^k / V_{i.kk}^i . \end{aligned}$$

Similarly

$$E(Z_{k.i}^2) = E\left[X_k(X_k - X_{k.1}^*)\right] = \frac{V_i^i}{V_{k.ii}^k} = \frac{V_i^i}{V_{i.kk}^i},$$

since

$$V_{k.ii}^k = V_{i.kk}^i .$$

We also have

$$\begin{aligned} E(Z_{i.k} Z_{k.i}) &= E(X_i - X_{i.k}^*)(X_k - X_{k.i}^*) \\ &= E[X_i(X_k - X_{k.i}^*)] - E[X_{i.k}^*(X_k - X_{k.i}^*)] \\ &= E[X_i(X_k - X_{k.i}^*)] - 0 . \end{aligned}$$

Indeed, $X_k - X_{k.i}^*$ is independent of $X_{i.k}^*$, which is a linear combination of the X_j.

We have thus

$$\begin{aligned} E(Z_{i.k} Z_{k.i}) &= E[X_i(X_k - X_{k.i}^*)] \\ &= E\left(X_i X_k + \sum_j \frac{V_{k.ii}^j}{V_{k.ii}^k} X_i X_j\right) = \sum_j \frac{V_{k.ii}^j}{V_{k.ii}^k} v_i^j . \end{aligned}$$

The sum $\sum_j V_{k.ii}^k v_i^j$ is the expansion with respect to the kth row of the determinant derived from V_{ii} by replacing its kth row $v_k^1, v_k^2, \ldots, v_k^k$ by $v_i^1, v_i^2, \ldots, v_i^n$. Let D_1 be this determinant. By performing $k - i - 1$ permutations on this kth row $v_i^1, v_i^2, \ldots, v_i^n$ we obtain the determinant D_2 derived from $|V|$ by removing its kth row and ith column. D_2 is therefore the minor V_k^i divided by $(-1)^{i+k}$, and we finally have

$$D_1 = V_k^i(-1)^{k-i-1-k-i} = -V_k^i .$$

We have thus

$$E(Z_{i.k}\, Z_{k.i}) = -V_k^i / V_{k.ii}^k .$$

Note here that

$$V_k^i = V_i^k \quad \text{and} \quad V_{k.ii}^k = V_{i.kk}^i .$$

We deduce from these relation that

$$\rho_{i.n-2}^k = \frac{E(Z_{i.k}\, Z_{k.i})}{\sqrt{E(Z_{i.k}^2)\, E(Z_{k.i}^2)}} = -\frac{V_i^k}{V_{k.ii}^k}\sqrt{\frac{V_{k.ii}^k}{V_i^i V_k^k}} = -\frac{V_i^k}{\sqrt{V_i^i\, V_k^k}}$$

that is,

$$\text{(X-13)} \qquad \boxed{\rho_{i.n-2}^k = -\frac{V_i^k}{\sqrt{V_i^i\, V_k^k}}} \; .$$

Note that these results can be expressed in terms of the correlation matrix $[R]$.

We have indeed $\quad |V| = |R|\, \sigma_1^2 \sigma_2^2 \ldots \sigma_n^2$

and $\quad V_i^k = R_i^k\, \sigma_1^2 \sigma_2^2 \ldots \sigma_{i-1}^2\, \sigma_{i+1} \ldots \sigma_{k-1}^2\, \sigma_{k+1}^2 \ldots \sigma_n^2 ,$

$$V_i^i = R_i^k\, \sigma_1^2 \sigma_2^2 \ldots \sigma_{i-1}^2\, \sigma_{i+1}^2 \ldots \sigma_n^2 .$$

We therefore find the following values for the regression coefficients:

$$\alpha_i^k = -\frac{V_i^k}{V_i^i} = -\frac{\sigma_i}{\sigma_k}\frac{R_i^k}{R_i^i}.$$

Similarly for the multiple correlation coefficient:

$$\rho_i = \sqrt{1 - \frac{|V|}{v_i^i V_i^i}} = \sqrt{1 - \frac{|R|}{R_i^i}}. \tag{X-14}$$

And finally for the partial correlation coefficient:

$$\rho_{i.n-2}^k = -\frac{V_i^k}{\sqrt{V_i^i V_k^k}} = -\frac{R_i^k}{\sqrt{R_i^i R_k^k}}. \tag{X-15}$$

We see the fundamental role of the matrices $[V]$ and $[R]$ in these questions of linear regression. They allow us to define all the coefficients we need.

We also note, before closing this section, that the concepts that arise in this way are of real interest only in the case where the law governing the set of variables $X_1, X_2, \ldots, X_n$ is normal.

II. EXPLICIT FORM OF THE LAPLACE-GAUSS LAW FOR n VARIABLES

As a second application of the techniques of matrix calculus to the calculus of probabilities, we now determine the explicit form of the non-degenerate Laplace-Gauss law for n variables and deduce its characteristic function and its covariance matrix.

The probability density of the Laplace-Gauss law for n variables can be expressed in its most general form as follows:

$$f(x_1, x_2, \ldots, x_n)\, dx_1\, dx_2 \ldots dx_n = K \exp[-\tfrac{1}{2} Q(x_1, x_2, \ldots, x_n)]\, dx_1\, dx_2 \ldots dx_n$$

where $Q(x_1, x_2, \ldots, x_n)$ is a positive definite or semidefinite quadratic form and K is a constant such that the integral of f over the whole space E_n is equal to 1. We suppose in the sequel that the quadratic form is positive definite.

In matrix notation, the expression above is written as follows:

$$f(X)\, dX = K \exp(-\tfrac{1}{2} X'[\Phi] X)\, dX ,$$

where X is a column vector:

$$X = \begin{bmatrix} x_1 \\ x_2 \\ \vdots \\ x_n \end{bmatrix}$$

and $[\Phi]$ is the matrix of the quadratic form Q. The symbol dX represents the differential element $dx_1\, dx_2 \ldots dx_n$.

The value of the constant K obviously depends on the quadratic form Q, and hence on $[\Phi]$, and we finally have

$$f(X)\, dX = K(\Phi) \exp(-\tfrac{1}{2} X'[\Phi] X)\, dX .$$

We now evaluate $K(\Phi)$ in order to determine the probability law completely. We first recall a result we will not prove here:

$$\overbrace{\int \cdots \int}^{n \text{ times}} \exp[-\tfrac{1}{2}(x_1^2 + x_2^2 + \cdots + x_n^2)]\, \mathrm{d}x_1\, \mathrm{d}x_2 \cdots \mathrm{d}x_n = (2\,\pi)^{n/2}$$

or in matrix notation

$$\int_{E_n} \exp(-\tfrac{1}{2} X'X)\, \mathrm{d}X = (2\,\pi)^{n/2}.$$

In order to evaluate $K(\Phi)$ we now perform a change of basis defined by a matrix $[B]$:

$$X = [B]\, X_B .$$

It follows that

$$X'[\Phi]\, X = X_B'[B]'\,[\Phi]\,[B]\, X_B$$

and therefore

$$\int_{E_n} K(\Phi) \exp(-\tfrac{1}{2} X'[\Phi]\, X)\, \mathrm{d}X =$$

$$= \int_{E_n} K(\Phi) \exp(-\tfrac{1}{2} X_B'[B']\,[\Phi]\,[B]\, X_B) \frac{D(X)}{D(X_B)} \mathrm{d}X_B$$

where $D(X)/D(X_B)$ is the Jacobian of the change of variables. Now

$$\frac{D(X)}{D(X_B)} = |\,B\,|,$$

which is obvious if we recall that the general term of $D(X)/D(X_B)$ is the partial derivative of the ith coordinate of the old system with respect to the kth coordinate of the new system. Since the change of variables is linear (see the formulas on page 58), this general term is b_i^k, that is, the general term of $[B]$.

We can therefore write

$$\int_{E_n} K(\Phi) \exp(-\tfrac{1}{2} X'X) \, \mathrm{d}X = \int_{E_n} K(\Phi) \exp(-\tfrac{1}{2} X'_B B'[\Phi] B X_B) \mid B \mid \mathrm{d}X_B .$$

We choose the change of basis so that $B'[\Phi] B = [I]$. which is possible since the matrix $[\Phi]$ is non-singular (the law is non-degenerate).

Looking at the determinants, we find

$$\mid B' \mid \mid \Phi \mid \mid B \mid = \mid B \mid^2 \mid \Phi \mid = 1 ,$$

that is

$$\mid B \mid = \frac{1}{\sqrt{\mid \Phi \mid}}$$

and therefore

$$\frac{K(\Phi)}{\sqrt{\mid \Phi \mid}} \int_{E_n} \exp(-\tfrac{1}{2} X'_B X_B) \, \mathrm{d}X_B = \frac{K(\Phi)}{\sqrt{\mid \Phi \mid}} (2\pi)^{n/2} = 1 .$$

We have thus

$$K(\Phi) = \frac{\sqrt{\mid \Phi \mid}}{(2\pi)^{n/2}} .$$

And the Laplace-Gauss law can finally be written in the following form:

$$f(X) \, \mathrm{d}X = \frac{\sqrt{\mid \Phi \mid}}{(2\pi)^{n/2}} \exp(-\tfrac{1}{2} X'[\Phi] X) \, \mathrm{d}X \qquad \text{(X-16)}$$

COVARIANCE MATRIX

The covariance between X_i and X_k can be defined as follows:

$$E(X_i X_k) = \int \int x_i x_k f(x_1, x_2, \ldots, x_n) \, dx_1 . dx_2 . \ldots . dx_n .$$

so that the covariance matrix can be defined as follows:

$$V(\Phi) = \frac{\sqrt{|\Phi|}}{(2\pi)^{n/2}} \int_{E_n} XX' \exp(-\tfrac{1}{2} X'[\Phi] X) \, dX .$$

In this relation, XX' represents a square matrix whose general term is precisely v_i^k.

We perform, as above, a change of basis defined by a matrix $[B]$ such that the matrix $[B]' [\Phi] [B]$ is the unit matrix. We have thus

$$V(\Phi) = \frac{\sqrt{|\Phi|}}{(2\pi)^{n/2}} \int_{E_n} [B] X_B X'_B [B'] \exp(-\tfrac{1}{2} X'_B [B'] [\Phi] [B] X_B) \; |B| \; dX_B$$

which can be written as follows, since the matrix $[B]' [\Phi] [B]$ is the unit matrix:

$$V(\Phi) = \sqrt{|\Phi|} \, |B| \, [B] \left[\int_{E_n} X_B X'_B \exp(-\tfrac{1}{2} X'_B X_B) \, dX \right] [B'] .$$

If we note that $|B|$ is equal to $1/\sqrt{|\Phi|}$ and the integral between brackets is simply $V(I)$, we have

$$V(\Phi) = [B] \, V(I) \, [B'] .$$

Now, $V(I)$ is the covariance matrix of the reduced centered law encountered earlier. All the covariances v_i^k are zero and the variances v_i^i are equal to 1. We have thus

$$V(I) = [I]$$

and therefore

$$V(\Phi) = [B] \, [B]' .$$

But the matrix $[B]$ is determined by the following condition:

$$[B]'[\Phi][B] = [I],$$

which implies

$$[B]' = [(\Phi B)]^{-1} = [B]^{-1}[\Phi]^{-1} \quad ,$$

and

$$[B][B]' = [B][B]^{-1}[\Phi]^{-1} = [\Phi]^{-1}.$$

And we finally get

(X-17) $$\boxed{V(\Phi) = [\Phi]^{-1}} \ .$$

*The covariance matrix of the non-degenerate normal law for **n** variables is the inverse of the matrix defining this normal law.*

CHARACTERISTIC FUNCTION OF THE NORMAL LAW FOR n VARIABLES

The characteristic function is defined as follows for a probability law with one variable:

$$\varphi_x(t) = E(e^{itX}) = \int e^{itx}\, dF(x).$$

For a probability law in n variables, the characteristic function is defined as follows:

$$\varphi_{x_1,x_2,\dots,x_n}(t_1, t_2, \dots, t_n) = E(\exp[i \sum_j t_j x_j])$$

that is, in matrix notation,

$$\varphi_X(T) = E(e^{iX'T}),$$

where

$$T = \begin{bmatrix} t_1 \\ t_2 \\ \cdot \\ \cdot \\ \cdot \\ t_n \end{bmatrix}$$

and thus, for the normal law for n variables:

$$\varphi_X(T) = \frac{\sqrt{|\Phi|}}{(2\pi)^{n/2}} \int_{E_n} e^{iX't} e^{-\frac{1}{2}X'[\Phi]X} dx .$$

We will not prove the following result which we will be using:

$$\frac{1}{(2\pi)^{n/2}} \int_{E_n} e^{iX'U} e^{-iX'X} dx = e^{-\frac{1}{2}U'U}.$$

It is easy to verify it, replacing the vector notation by the usual algebraic notation and thus transforming the integral into a product of n simple integrals, each equal to

$$\frac{1}{\sqrt{2\pi}} \int_{-\infty}^{+\infty} e^{iux} e^{-x^2/2} dx = e^{-u^2/2} .$$

In the characteristic function

$$\varphi_X(T) = \frac{\sqrt{[\Phi]}}{(2\pi)^{n/2}} \int_{E_n} e^{iX'T} e^{-\frac{1}{2}X'[\Phi]X} dX$$

we then perform, as previously, a change of basis defined by the relation

$$X = [B]\, X_B ,$$

where $[B]$ is defined by the condition

$$[B]'\,[\Phi]\,[B] = [I] ,$$

which allows us to write

$$\varphi_X(T) = \frac{\sqrt{|\Phi|}}{(2\pi)^{n/2}} \int_{E_n} \exp(\mathrm{i}\, X'_B[B]'\, T) \exp(-\tfrac{1}{2} X'_B X_B)\, |B|\, \mathrm{d}X_B .$$

In this expression $[B]$ is equal to $1/\sqrt{|\Phi|}$, so that the integral can be expressed as follows:

$$\varphi_X(T) = \frac{1}{(2\pi)^{n/2}} \int_{E_n} \exp(\mathrm{i}\, X'_B[B]'\, T) \exp(-\tfrac{1}{2} X'_B X_B)\, |B|\, \mathrm{d}X_B .$$

It is then sufficient to set $[B]'\, T = U$ in order to get back to the expression written above. We get:

$$\varphi_X(T) = \mathrm{e}^{-\frac{1}{2}U'U} = \mathrm{e}^{-\frac{1}{2}(B'T')B'T} = \mathrm{e}^{-\frac{1}{2}T'[B][B]'T}.$$

Now, we have seen earlier that

$$[B]\,[B]' = [\Phi]^{-1} .$$

It therefore finally follows that

(X-18) $$\boxed{\varphi_X(T) = \mathrm{e}^{-\frac{1}{2}T'[\Phi]^{-1}T}}$$

or, if we want to use the covariance matrix $[V]$:

(X-19) $$\boxed{\varphi_X(T) = \mathrm{e}^{-\frac{1}{2}T'[V]T}}$$

(X-19) *There is therefore a simple expression for the characteristic function of the Laplace-Gauss law for* n *variables in terms of the covariance matrix.*

III. STOCHASTIC MATRICES

We now examine a probability problem which will give us the opportunity to make a thorough study of certain techniques of linear algebra.

Consider a random phenomenon which changes with time. This phenomenon is defined at equidistant discrete times $t_0, t_1, t_2, \ldots, t_n, \ldots$ At each of the moments when it is defined, this phenomenon, which is denoted $E(t)$, can assume one of the following values:

$$E_1, E_2, \ldots, E_m .$$

The phenomenon is completely defined statistically by the probabilities of the various events possible in the course of its evolution in time during periods of arbitrary length.

We make two hypotheses:

a) The probability of obtaining a certain result E_i at time t_j, knowing that results $\ldots, E_{k_1}, E_{k_1+1}, \ldots, E_{k_{j-1}}$ have been obtained, depends on the result $E_{k_{j-i}}$ alone and not on earlier results.

b) This probability is independent of the time t_j.

We then say that the random phenomenon under consideration is a simple stationary Markov chain.

This Markov chain is defined by the transition probabilities p_i^k:

$$p_i^k = P[E(t_j) = E_i/E(t_{j-1}) = E_k]$$

and by the initial probabilities

$$a_i = P[E(t_0) = E_i] .$$

The numbers p_i^k and a_i cannot be arbitrary since they are probabilities; they are non-negative and satisfy the following relations:

$$\sum_i a_i = 1 , \qquad \sum_i p_i^k = 1 , \qquad k = 1, 2, \ldots, n .$$

The first relation expresses the fact that a result has been obtained certainly at the initial moment, the second, that if a result E_k has been obtained at time $j-1$, a result is obtained certainly at time j.

We are going to show that these relations are the only ones that need be imposed in order to ensure that the a_i and the p_i^k represent a random phenomenon.

We now look for the probability $\mathcal{T}$ of a sequence of results $E_{j_0}, E_{j_1}, E_{j_2}, \ldots, E_{j_{m-1}}, E_{j_m}$. We find

$$\mathcal{T} = a_{j_0}\, p_{j_1}^{j_0}\, p_{j_2}^{j_1} \ldots p_{j_m}^{j_{m-1}}\,.$$

We must verify that $\mathcal{T}$ lies between 0 and 1 and that the sum of all probabilities $\mathcal{T}$ corresponding to the various possible events is equal to 1. Since a_i and p_i^j are probabilities, it follows that $\mathcal{T}$ lies between 0 and 1. In order to verify that $\Sigma\mathcal{T}$ is equal to 1, we first sum over the index j_m. We find:

$$\sum_{j_m} \mathcal{T} = a_{j_0}\, p_{j_1}^{j_0}\, p_{j_2}^{j_1} \ldots p_{j_{m-1}}^{j_{m-2}} \sum_{j_m} p_{j_m}^{j_{m-1}}\,.$$

The sum in the right-hand side is equal to 1 in view of the relations satisfied by the p_i^j. We then sum over j_{m-1}, and so on. We finally get the equality

$$\sum \mathcal{T} = \sum_i a_i = 1\,.$$

The a_i and p_i^j therefore form a set of elements which completely define the Markov chain under consideration.

We now consider the following matrix:

$$[P] = \begin{bmatrix} p_1^1 & p_1^2 & \dots & p_1^n \\ p_2^1 & p_2^2 & \dots & p_2^n \\ \cdot & \cdot & \cdot & \cdot \\ p_n^1 & p_n^2 & \dots & p_n^n \end{bmatrix}.$$

It is the matrix of the Markov chain under consideration. This matrix is not arbitrary since all its elements are non-negative and the sum of the elements in each column is equal to 1. *Such a matrix is called a stochastic matrix.*

Stochastic matrices will allow us to study the properties of Markov chains in a very convenient way, as we will now see.

PROPERTIES OF STOCHASTIC MATRICES

If two matrices $[A]$ *and* $[B]$ *are stochastic, so is the matrix* $[p]\,[A] + [q]\,[B]$, *where* $[p]$ *is the scalar matrix whose diagonal elements are equal to* p *and* $[q]$ *is the scalar matrix whose diagonal elements are equal to* q, *where* $p + q = 1$.

We have indeed

$$[p]\,[A] = \begin{bmatrix} pa_1^1 & pa_1^2 & \dots & pa_1^n \\ pa_2^1 & pa_2^2 & \dots & pa_2^n \\ \cdot & \cdot & \cdot & \cdot \\ pa_n^1 & pa_n^2 & \dots & pa_n^n \end{bmatrix}$$

$$[q]\,[B] = \begin{bmatrix} qb_1^1 & qb_1^2 & \dots & qb_n^n \\ qb_2^1 & qb_2^2 & \dots & qb_2^n \\ \cdot & \cdot & \cdot & \cdot \\ qb_n^1 & qb_n^2 & \dots & qb_n^n \end{bmatrix}$$

and therefore

$$[p][A] + [q][B] = \begin{bmatrix} pa_1^1 + qb_1^1 & pa_1^2 + qb_1^2 & \dots & pa_1^n + qb_1^n \\ pa_2^1 + qb_2^1 & pa_2^2 + qb_2^2 & \dots & pa_2^n + qb_2^n \\ \dots & \dots & \dots & \dots \\ pa_n^1 + qb_n^1 & pa_n^2 + qb_n^2 & \dots & pa_n^n + qb_n^n \end{bmatrix}.$$

All the elements of the matrix $[p][A] + [q][B]$ are positive or zero; if on the other hand we sum the elements in a column, we get

$$\sum_i (pa_i^k + qb_i^k) = \sum_i pa_i^k + qb_i^k = p \sum_i a_i^k + q \sum_i b_i^k = p + q = 1 .$$

The matrix $[p][A] + [q][B]$ is therefore stochastic. Q.E.D.

(X-20) *If the matrices* $[A]$ *and* $[B]$ *are stochastic, so are the matrices* $[A][B]$ *and* $[B][A]$.

Let $[P]$ be the matrix $[A][B]$. We have

$$p_i^k = \sum_l a_i^l \, b_l^k$$

and therefore

$$\sum_i p_i^k = \sum_i \sum_l a_i^l \, b_l^k .$$

Reversing the order of summation -- which can always be done -- we find

$$\sum_i p_i^k = \sum_l \sum_i a_i^l \, b_l^k = \sum_l b_l^k \sum_i a_i^l = \sum_l b_l^k = 1 .$$

The matrix $[A][B]$ is therefore stochastic. We could prove in the same way that the matrix $[B][A]$ is also stochastic.

We now try to determine the probabilities $p_k(t)$ of the events E_k at each instant t.

The probability of E_k at time t_0 is a_k, that is,

$$p_k(t_0) = a_k .$$

We denote by $V(0)$ the vector of n-dimensional space whose components are $a_1, a_2, \dots, a_n$. It is the *probability vector at the initial moment.* In general, the probability vector at time t, denoted by $V(t)$, is defined as the vector whose components are the probabilities of the results $E_1, E_2, \dots, E_n$ at time t.

We now look for the probability of obtaining E_i at time 1. Applying the total probability and compound probability formulas, we find

$$p_i(1) = a_1\, p_i^1 + a_2\, p_i^2 + \dots + a_n\, p_i^n .$$

In matrix notation, we find therefore

$$V(1) = [P]\, V(0) .$$

More generally

$$V(t) = [P]\, V(t-1) = [P]^t\, V(0)$$

or, if we wish to emphasize that t is discrete, we can write

$$V(m) = [P]\, V(m-1) = [P]^m\, V(0) .$$

All the properties of simple Markov chains can be deduced from the properties of the matrix $[P]$.

We first study some properties of probability vectors.

(X-21) *The sum of the components of a probability vector is equal to* 1.

This follows from the definition since the components represent all the probabilities of a probability law.

(X-22) *The vector* $[P]X$, *where* $[P]$ *is a stochastic matrix and* X *is a probability vector, is a probability vector.*

We must show that the sum of the components of $[P]X$ is equal to 1.

Let y_i be the components of $[P]X$. Then

$$y_i = \sum_j p_i^j x_j$$

and

$$\sum_i y_i = \sum_i \sum_j p_i^j x_j \sum_j \sum_i p_i^j x_j = \sum_j x_j \sum_i p_i^j .$$

Now

$$\sum_i p_i^j = 1 .$$

Hence

$$\sum_i y_i = \sum_j x_j = 1 ,$$

since X is a probability vector.

We have thus verified that we have actually defined a probability law by means of the matrix $[P]$, since at any time the sum of the probabilities of $E_1, E_2, \ldots, E_n$ is indeed equal to 1.

We can deduce another definition of stochastic matrices:

(X-23) *A matrix* $[P]$ *is stochastic if and only if the vector* $[P]X$ *is a probability vector whenever the vector* X *is a probability vector.*

We have seen that the condition is necessary; we now show that it is sufficient.

Let y_i be the components of the vector $[P]X$. Then

$$y_i = \sum_i p_i^j x_j$$

and, by hypothesis,

$$\sum_i y_i = \sum_i \sum_j p_i^j x_j = 1 .$$

Now, we can also write

$$\sum_i \sum_j p_i^j x_j = \sum_j \sum_i p_i^j x_j = \sum_j x_j \sum_i p_i^j = 1 .$$

And this is true for any probability vector X. It is therefore true for the vector e_j all of whose coordinates are zero except the jth, e_j^j, which is equal to 1. Thus,

$$\sum_j x_j \sum_i p_i^j = e_j \sum_i p_i^j = \sum_i p_i^j = 1 .$$

Taking successively $j = 1, 2, \ldots, n$, we see that all the column sums are equal to 1. The matrix $[P]$ is therefore stochastic.

Theorem.

(X-24) *All the eigenvalues of a stochastic matrix have modulus at most equal to* 1, *and one of them is equal to* 1.

Consider an eigenvector X and the associated eigenvalue λ. We have

$$[P]X = Y = \lambda X .$$

Considering the transpose vectors, we find

$$X'[P]' = Y' = \lambda X'$$

where on the one hand

$$y_i' = \sum_j x_i p_j^i \text{ and, on the other, } y_i' = \lambda x_i .$$

We denote by $x_{\max}$ and $y'_{\max}$ the coordinates with greatest modulus of the vectors X' and Y', respectively.

The first relation above yields

$$| y_i' | \leqslant | x_{\max} | \sum_j p_j^i = | x_{\max} | \quad \text{for any } i$$

since the matrix $[P]$ is stochastic. In particular

$$| y'_{\max} | \leqslant | x_{\max} | ;$$

but the second relation yields

$$| y'_{\max} | = | \lambda | | x_{\max} | ,$$

which implies

$$| \lambda | | x_{\max} | \leqslant | x_{\max} |$$

that is,

$$| \lambda | \leqslant 1 .$$

and this is true for any eigenvalue.

We now show that 1 is an eigenvalue.

We no longer consider the matrix $[P]$ but its transpose $[P]'$. We show that the vector K all of whose coordinates k_i are equal to 1 is an eigenvector of this matrix and that it is associated with the eigenvalue 1.

The coordinates of the vector $[P']K$ are

$$\pi_i = \sum_j p_j^i k_j = \sum p_j^i = 1 \quad \text{for any } i ,$$

and hence

$$[P'] K = K$$

which shows that K is an eigenvector associated with the eigenvalue 1. Now, the matrices $[P]$ and $[P']$ have the same eigenvalues. It follows that $[P]$ admits 1 as an eigenvalue.

We now prove the following fundamental theorem.

Ergodic Theorem.

(X-24) *In a simple Markov chain where the stochastic matrix $[P]$ has strictly positive elements (no zero elements), the probability vector converges, as t increases indefinitely, to a vector V which is independent of the initial probability vector $V(0)$.*

As defined earlier,

$$V(t) = [P]^t\, V(0) .$$

We must show that this quantity converges to a vector V_L which is independent of $V(0)$ as t increases indefinitely. Since t can only assume integer values, we replace t by m and consider

$$V(m) = [P]^m\, V(0) = [P]^m\, V_0 .$$

We first prove a preliminary lemma:

(X-25) *In an n-dimensional space (where n is finite), if the scalar product $X_m' B$ converges for any B as m increases indefinitely, then the vector X_m converges to a limit as m increases indefinitely.*

We need only show that each coordinate of X_m converges to a limit as m increases indefinitely. This is why the condition that m is finite is essential in this case since when m is infinite each coordinate can converge without the vector itself being convergent.

We successively take the various vectors e_1 as our vector B. We immediately find:

$$X'_m e_i = x_i^m \to x_i \quad \text{as} \quad m \to \infty .$$

Hence

$$X_m \to X , \quad \text{where} \quad X = (x_1, x_2, \ldots, x_n) .$$

X_m therefore converges to a well-determined vector as m increases indefinitely.

Having established this we now consider the following scalar product:

$$B'_0[P]^m V_0 ,$$

where B_0 is an arbitrary vector of E_n. From the definition of the transpose of the matrix $[P]$, we have

$$B'_0[P]^m V_0 = V'_0[P^m]' B_0 .$$

We now study the vector $[P^m]'B_0$ directly. We have indeed considered the scalar product $B'_0[P^m] V_0$ only to introduce the transpose matrix $[P]'$ in which the *row sums* (instead of the column sums) are equal to 1.

We set

$$[P^m]' B_0 = B^m .$$

And we therefore find

$$B^{m+1} = [P^{m+1}]' B_0 = [P'] [P^m]' B_0 = [P]' B^m$$

and thus, passing to coordinates,

$$b_i^{m+1} = \sum_j (p_i^j)' b_j^m = \sum_j p_j^i b_j^m .$$

If we denote by $b_{\max}^m$ the greatest of the b_i^m and by $b_{\min}^m$ the smallest, we immediately find

$$b_i^{m+1} \leqslant \sum_j p_j^i b_{\max}^m = b_{\max}^m \sum_j p_j^i = b_{\max}^m$$

and similarly

$$b_i^{m+1} \geqslant b_{\min}^m .$$

These relations are valid in particular for $b_{\max}^{m+1}$ and $b_{\min}^{m+1}$, which allows us to write:

$$b_{\max}^{m+1} \leqslant b_{\max}^m , \qquad \text{with} \quad b_{\max}^{m+1} \geqslant 0 ,$$

$$b_{\min}^{m+1} \geqslant b_{\min}^m , \qquad \text{with} \quad b_{\min}^{m+1} \leqslant 1 .$$

The first sequence is non-increasing and bounded from below, and has therefore a limit $b_{\max}$; the second is non-decreasing and bounded from above, and is therefore convergent also and $b_{\min}^{m+1}$ has a limit $b_{\min}$.

We now show that $b_{\min}$ and $b_{\max}$ are equal.

To do this, let d be the smallest of the probabilities p_i^j . In view of our fundamental hypothesis that $[P]$ is strictly positive, d must be positive.

We can therefore write

$$b_i^{m+1} = \sum_j p_j^i b_j^m = p_{j'}^i b_{\min}^m + d_i b_{j''}^m + \sum_{\substack{j \neq j' \\ j \neq j''}} p_j^i b_j^m$$

emphasizing the term in $b_{\min}^m$ and that in d_i (where d_i is the smallest probability in the ith row).

Now, by hypothesis,

$$p^i_{j'} \geqslant d_i , \qquad b^m_{j''} - b^m_{\min} \geqslant 0 .$$

Hence

$$p^i_{j'}(b^m_{j''} - b^m_{\min}) \geqslant d_i(b^m_{j''} - b^m_{\min})$$

or

$$p^i_{j'}\, b^m_{j''} - p^i_{j'}\, b^m_{\min} \geqslant d_i\, b^m_{j''} - d_i\, b^m_{\min} .$$

We can therefore write

$$b^{m+1}_i \leqslant p^i_{j'}\, b^m_{j''} + d_i\, b^m_{\min} + \sum_{\substack{j \neq j' \\ j \neq j''}} p^i_j\, b^m_j$$

and a fortiori

$$b^{m+1}_i \leqslant d_i\, b^m_{\min} + b^m_{\max} \left(p^i_{j'} + \sum_{\substack{j \neq j' \\ j \neq j''}} p^i_j \right) .$$

Now

$$p^i_{j'} + \sum_{\substack{j \neq j' \\ j \neq j''}} p^i_j \geqslant 1 - d_i$$

hence

$$b^{m+1}_i \leqslant d_i\, b^m_{\min} + (1 - d_i)\, b^m_{\max} .$$

Since

$$d_i - d \geqslant 0 \qquad \text{and} \qquad b^m_{\max} - b^m_{\min} \geqslant 0 ,$$

we can deduce that

$$(d_i - d)\,(b^m_{\max} - b^m_{\min}) \geqslant 0 ,$$

that is,

$$d_i\, b^m_{\max} - d_i\, b^m_{\min} - db^m_{\max} + db^m_{\min} \geqslant 0$$

or, adding $b_{\max}$ to each side and transferring the terms in d_i to the right-hand side:

$$db^m_{\min} + (1 - d)\, b^m_{\max} \geqslant d_i\, b^m_{\min} + (1 - d_i)\, b^m_{\max}\,;$$

we finally get

$$b^{m+1}_1 \leqslant db^m_{\min} + (1 - d)\, b^m_{\min}$$

and in particular

$$b^{m+1}_{\max} \leqslant db^m_{\min} + (1 - d)\, b^m_{\max}\,,$$

$$b^{m+1}_{\min} \geqslant db^m_{\max} + (1 - d)\, b^m_{\min}\,.$$

Therefore

$$(b^{m+1}_{\max} - b^{m+1}_{\min}) \leqslant (1 - 2\,d)\,(b^m_{\max} - b^m_{\min})$$

and finally

$$b^{m+1}_{\max} - b^{m+1}_{\min} \leqslant (1 - 2\,d)\,(b^m_{\max} - b^m_{\min})\,.$$

Repeating this process yields

$$b^{m+1}_{\max} - b^{m+1}_{\min} \leqslant (1 - 2\,d)^n\,(b^0_{\max} - b^0_{\min})\,.$$

It follows that the two quantities $b^m_{\max}$ and $b^m_{\min}$ converge to the same limit b as m increases indefinitely and that the vector $[P^m]'\,B_0$ converges to the vector B all of whose components are equal to b. It follows that

$$\lim_{m\to\infty} B'_0[P]^m\, V_0\, B'\, V_0 = (V_0, B) = \sum_i v^0_i\, b_i = b \sum_i v^0_i = b\,,$$

since the vector V_0 is a stochastic vector.

The limit $[P]^m\, V_0$ is therefore independent of the

initial vector V_0 and the theorem is proved.

We can immediately verify that the limit vector V_L is an eigenvector of the matrix $[P]$ and that it is associated with the eigenvalue 1.

We have indeed

$$\lim [P]^m V_0 = \lim [P] [P]^{m-1} V_0 = [P] \lim [P]^{m-1} = [P] V_L = V_L .$$

If all the roots of the characteristic equation of the transition matrix $[P]$ are distinct, there is a far simpler proof for the above result.

We must prove that the vector $[P]^m V_0$ converges to a fixed vector independent of V_0 as m increases indefinitely.

We perform a change of basis in E_n, taking as our new basis vectors the eigenvectors of the matrix $[P]$ (there are n of them since the eigenvalues of $[P]$ are all distinct). This change of basis is orthogonal but it does not preserve the stochastic properties of the matrix $[P]$ and the vector V_0.

Denoting the matrix of the change of basis $[C]$, we get

$$V_{0c} = [C'] V_0 = \alpha_1 \Lambda_1 + \alpha_2 \Lambda_2 + \cdots + \alpha_n \Lambda_n .$$

The matrix $[P_2]$ corresponding to $[P]$ is $[C'][P][C]$ and it has the same eigenvalues as $[P]$. The eigenvalue of $[P_2]$ with greatest modulus is therefore 1 and we assume that the eigenvector Λ_1 corresponds to this eigenvalue 1.

We can therefore write

$$
\begin{aligned}
[P_c]\, V_{0c} &= [P_c]\,[\alpha_1 \Lambda_1 + \alpha_2 \Lambda_2 + \cdots + \alpha_n \Lambda_n] \\
&= \alpha_1 \Lambda_1 + \lambda_2 \alpha_2 \Lambda_2 + \cdots + \lambda_n \alpha_n \Lambda_n = V_{1c} \\
[P_c]^2\, V_{0c} &= [P_c]\, V_{1c} = \alpha_1 \Lambda_1 + \lambda_2^2 \alpha_2 \Lambda_2 + \cdots + \lambda_n^2 \alpha_n \Lambda_n = V_{2c} \\
[P_c]^m\, V_{0c} &= \alpha_1 \Lambda_1 + \lambda_2^m \alpha_2 \Lambda_2 + \cdots + \lambda_n^m \alpha_n \Lambda_n = V_{mc}\,.
\end{aligned}
$$

Since the eigenvalues $\lambda_2, \ldots, \lambda_n$ all have modulus smaller than 1, the powers λ_j^m tend to 0 as n increases indefinitely and we finally get

$$\lim_{m\to\infty} P_c V_{0c} = \alpha_1 \Lambda_1 = V_c\,.$$

We then get back to the first basis by setting

$$V = [C]\, V_c\,.$$

Now, since V_c is an eigenvector of $[P_c]$ associated with the eigenvalue 1, it follows that

$$[P_c]\, V_c = V_c\,.$$

Introducing $[P]$ and V which satisfy the following relations

$$V_c = [C']\, V \qquad \text{and} \qquad [P_c] = [C']\,[P]\,[C]$$

we get

$$[C']\,[P]\,[C]\,[C']\, V = [C']\, V,$$

which implies, since the change of basis is orthogonal,

$$[P]\, V = V\,.$$

In other words, the vector V is also an eigenvector of the matrix $[P]$ and is also associated with the eigenvalue 1.

It follows that

$$V = \lim_{m \to \infty} [P]^m V_0 .$$

Denoting this limit by V', we find

$$\lim_{m \to \infty} [P]^m V_0 = [P] \lim_{m \to \infty} [P^{m-1}] V_0$$

or

$$V' = PV' .$$

Now, the relation

$$V = [C] V_c$$

shows that V is a stochastic vector.

Indeed, for any m,

$$[C] V_{mc} = V_m = [P]^m V_0 ,$$

that is, for any m, every vector $[C] V_{mc}$ is a stochastic vector; and the same is therefore true of V.

APPLICATION

Consider the simple Markov chain defined by the transition matrix $[P]$ and the initial probability vector V_0, where

$$P = \begin{bmatrix} 0.3 & 0.2 & 0.5 \\ 0.5 & 0.4 & 0.3 \\ 0.2 & 0.4 & 0.2 \end{bmatrix}, \qquad V_0 = \begin{bmatrix} 1 \\ 0 \\ 0 \end{bmatrix}.$$

We find successively

$$V_1 = \begin{bmatrix} 0.3 \\ 0.5 \\ 0.2 \end{bmatrix}, \quad V_2 = \begin{bmatrix} 0.29 \\ 0.41 \\ 0.30 \end{bmatrix}, \quad V_3 = \begin{bmatrix} 0.319 \\ 0.399 \\ 0.282 \end{bmatrix}, \quad V_4 = \begin{bmatrix} 0.316\,5 \\ 0.403\,7 \\ 0.279\,8 \end{bmatrix},$$

$$V_5 = \begin{vmatrix} 0.315\,59 \\ 0.403\,67 \\ 0.280\,74 \end{vmatrix}, \quad V_6 = \begin{vmatrix} 0.315\,791 \\ 0.403\,485 \\ 0.280\,724 \end{vmatrix}, \quad V_7 = \begin{vmatrix} 0.315\,796\,3 \\ 0.403\,506\,7 \\ 0.280\,697\,0 \end{vmatrix}.$$

We see that the convergence is quite rapid, since we have already obtained a probability vector to four significant figures in as few as seven steps.

We therefore also verify that the characteristic equation of the matrix $[P]$ admits the root 1 and that the vector $\boldsymbol{V}$, that is, "almost" the vector $\boldsymbol{V}_7$, is an eigenvector of $[P]$.

To verify that the matrix $[P]$ admits the eigenvalue 1, we need only write its characteristic polynomial.

$$|P - \lambda_1| = \begin{vmatrix} 0.3-\lambda & 0.2 & 0.5 \\ 0.5 & 0.4-\lambda & 0.3 \\ 0.2 & 0.4 & 0.2-\lambda \end{vmatrix}$$

that is

$$|P - \lambda I| = -\lambda^3 + 0.900\,\lambda^2 + 0.060\,\lambda + 0.040\,.$$

And we readily see that it can be written as follows:

$$|P - \lambda I| = -(\lambda - 1)(\lambda^2 + 0.10\,\lambda + 0.04)\,.$$

The polynomial $\lambda^2 + 0.10\ \lambda + 0.04$ admits two imaginary roots with modulus equal to 0.180, that is, smaller than 1, in agreement with Theorem (X-24).

BIBLIOGRAPHY

J. ACHER, *Algèbre linéaire et programmation linéaire*, Dunod.

A.C. AITKEN, *Determinants and matrices*, Oliver and Boyd, Ltd.

R.C. BELLMANN, *Introduction to matrix analysis*, McGraw-Hill Book Company.

H. CRAMER, *Mathematical methods of statistics*, Princeton University Press.

D. DUGUÉ, *Traité de statistique théorique et appliquée*, Masson et Cie.

D.T. FINKBEINER, *Introduction to matrices and linear transformations*, W.H. Freeman and Company.

F.R. GANTMACHER, *The theory of matrices* (Vols. 1 and 2), Chelsea Publishing Company.

A. LICHNEROWICZ, *Algèbre et analyse linéaire*, Masson et Cie.

R.M. THRALL and L. TORNHEIM, *Vector spaces and matrices*, John Wiley and Sons, Inc.

H.W. TURNBULL, *The theory of determinants, matrices and invariants*, Blackie and Son, Ltd.

1. Show that the union of two vector subspaces of E_n need not be a vector space.

2. Show that the set of free vectors of the usual three-dimensional euclidean space is a vector space as defined in Chapter 1.

3. Find the sum of the following matrices:

$$[A] = \begin{bmatrix} 3 & 1 & 5 & 0 \\ 7 & 2 & 3 & 1 \\ 1 & 8 & 9 & 5 \end{bmatrix} \text{ and } [B] = \begin{bmatrix} 0 & 5 & 8 & 2 \\ 1 & 1 & 3 & 4 \\ 7 & 9 & 0 & 2 \end{bmatrix}.$$

4. Find the sum of the following matrices:

$$[C] = \begin{bmatrix} 12 & 4 & 1 & 0 & 5 \\ 3 & 7 & 14 & 8 & 2 \end{bmatrix}$$

$$[D] = \begin{bmatrix} 9 & 2 & 17 & 11 & 8 \\ 0 & 15 & 9 & 6 & 12 \end{bmatrix}.$$

5. The matrices $[A]$, $[B]$, $[C]$ and $[D]$ of the last two exercises represent linear mappings. State the dimensions of the corresponding vector spaces.

6. Find the product of the following matrices:

$$E = \begin{bmatrix} 6 & 2 \\ 1 & 0 \\ 9 & \\ 7 & 3 \end{bmatrix} \text{ and } F = \begin{bmatrix} 13 & 7 & 2 & 1 & 5 \\ 2 & 0 & 9 & 2 & 1 \end{bmatrix}.$$

What does this product represent?

7. Consider the following matrices:

$$[G] = \begin{bmatrix} 9 \\ 5 \\ 3 \\ 7 \end{bmatrix} \text{ and } [M] = [0 \quad 3 \quad 3 \quad 1] .$$

Compute [*G*] [*H*] and [*H*] [*G*]. What do these products represent?

8. Consider two vector spaces E_n and E_m and a mapping f of E_n into E_m defined as follows:

$$f(X) = \alpha_1(X)\, U_1 + \alpha_2(X)\, U_2 + \cdots + \alpha_p(X)\, U_p ,$$

where $\alpha_1(X), \alpha_2(X), \ldots, \alpha_p(X)$ are scalar functions of *X*, and $U_1, U_2, \ldots, U_p$ are arbitrary but fixed vectors of E_m $(p < m)$.
Is f a linear mapping? If not, state conditions which make it linear. In that case, what is the dimension of its domain? Find the form of its matrix.

9. Given an arbitrary rectangular matrix [*A*], show that the matrix [*A*] [*A'*] is symmetric and has the same number of rows as the matrix [*A*].

10. Given the two matrices

$$[A] = \begin{bmatrix} 12 & 5 & 2 \\ 0 & 4 & 1 \\ 3 & 2 & 6 \end{bmatrix} \quad [B] = \begin{bmatrix} 1 & 2 & 3 \\ 4 & 5 & 6 \\ 7 & 8 & 9 \end{bmatrix}$$

compute the products

$$[A]\,[B] \quad \text{and} \quad [B]\,[A]$$

and verify that they are not equal.

11. Show that the product of two diagonal matrices is a diagonal matrix.

12. Given an arbitrary square matrix $[A]$, show that the matrix $[A] + [A']$ is symmetric and the matrix $[A] - [A']$ skew-symmetric.

13. Given two square $n \times n$ matrices $[A]$ and $[B]$ of ranks r and s, respectively, show that the rank of the product matrix $[A]\,[B]$ is a number q between $r + s - n$ and the smaller of the numbers r and s.

14. Show that the value of a determinant is unchanged when a linear combination of some of the columns is added to another column.

15. Show that the same holds for addition of a linear combination of rows to another row.

16. Use Exercises 14 and 15 to find a method for replacing a determinant of order n by a determinant of order $n - 1$.

17. Use the above procedure as many times as necessary to compute the following determinant in a simple way:

$$\begin{vmatrix} 3 & 1 & 7 & 2 & 4 \\ 5 & 2 & 2 & 4 & 3 \\ 7 & 3 & 5 & 8 & 2 \\ 1 & 3 & 4 & 9 & 5 \\ 3 & 1 & 2 & 7 & 3 \end{vmatrix}.$$

18. The same question for the following determinant:

$$\begin{vmatrix} 3 & 9 & 1 & 5 & 2 & 7 \\ 2 & 0 & 5 & 4 & 8 & 1 \\ 6 & 1 & 5 & 3 & 9 & 4 \\ 8 & 4 & 0 & 7 & 3 & 3 \\ 5 & 8 & 2 & 0 & 7 & 4 \\ 3 & 5 & 1 & 6 & 2 & 8 \end{vmatrix}.$$

19. Verify the following relation directly:

$$|AB| = |A||B|,$$

for the matrices

$$|A| = \begin{vmatrix} 3 & 2 & 5 \\ 9 & 7 & 2 \\ 5 & 3 & 4 \end{vmatrix} \qquad |B| = \begin{vmatrix} 2 & 8 & 3 \\ 1 & 5 & 7 \\ 4 & 2 & 1 \end{vmatrix}.$$

20. Find the conjugate permutations of the following permutations:

4 1 3 5 7 2 9 8 6
8 3 1 2 4 9 7 5 6
3 7 9 5 2 4 1 6 8
5 7 2 3 1 9 4 6 8
1 5 2 9 3 6 7 4 8
7 6 3 9 5 1 4 8 2.

21. Count the numbers of inversions in the permutations of Exercise 20, and verify directly that the conjugate permutations contain the same numbers of inversions.

22. Find the maximum possible number of inversions in a permutation of the first n integers. Find the corresponding permutations. Check your formula for permutations of the first 9 integers.

23. Is the following system of linear equations a Cramer system?

$$\begin{aligned} x + y + z &= 1 \\ x + y &= 2 \\ y + z &= 3\,. \end{aligned}$$

If so, use Cramer's formulas to solve it. Otherwise, find the solutions, if they exist.

24. The same for the following system:

$$\begin{aligned} x + y + z &= 1 \\ x + 2z &= 2 \\ y - z &= -1\,. \end{aligned}$$

25. For what values of α is the following system a Cramer system?

$$\begin{aligned} \alpha x + y - z &= 1\,, \\ x - \alpha y + z &= 2\alpha\,, \\ -x + y + \alpha z &= 1\,. \end{aligned}$$

Does this system have solutions when it is degenerate?

26. Consider the following system:

$$\begin{aligned} \alpha x + y + z &= A\,, \\ x + \alpha y + z &= B\,, \\ x + y + \alpha z &= C\,, \\ x + y + z &= 1\,. \end{aligned}$$

Find the order of the system for different (real) values of α. In each case, indicate the relations that must be satisfied by the parameters A, B and C in order that the system have a solution. Find the corresponding solution.

27. Given the bilinear function f defined by the matrix

I

$$[M] = \begin{bmatrix} -1 & 3 & 0 \\ 2 & 5 & 1 \\ 4 & 9 & 2 \end{bmatrix}$$

verify the invariance of $f(X, Y)$ directly when

$$X = (1, 2, 3)$$
$$Y = (5, 1, 0)$$

for the change of basis in E_3 defined by the following matrix:

$$[B] = \begin{bmatrix} 2 & 9 & 5 \\ 1 & 3 & 2 \\ 4 & 1 & 7 \end{bmatrix}.$$

28. It was shown in the text that the matrix of a bilinear function need not undergo the same transformation as the matrix of a linear mapping under a change of basis. Show that the two transformations are equivalent if and only if the change of basis is orthogonal.

29. Consider a vector V in the vector space E_n, and a non-singular $n \times n$ matrix $[M]$. With each vector X of E_n, associate a vector Y defined by

$$Y = V'[M]\,V = f_V(X).$$

What kind of mapping is f_V?
Consider the set of mappings f_V when V ranges over E_n. What can you say about this set? Is it a vector space? If so, find its dimension.

30. Do the vectors

$$X_1 = (4, 0, 3),$$
$$X_2 = (3, 3, 5),$$
$$X_3 = (1, 4, 7),$$

constitute a basis of E_3?
If so, use the Schmidt method to convert them into an orthonormal basis of E_3.

31. The same problem in the space E_4, with the following vectors:

$$\begin{aligned} A &= (1, 3, 5, 2)\,, \\ B &= (5, 0, 1, 3)\,, \\ C &= (9, 5, 0, 2)\,, \\ D &= (4, 2, 1, 3)\,. \end{aligned}$$

32. Given two vector subspaces $\mathcal{E}_1$ and $\mathcal{E}_2$ of E_n, and their orthogonal complements $\bar{\mathcal{E}}_1$ and $\bar{\mathcal{E}}_2$, respectively, prove the following relation:

$$\overline{\mathcal{E}_1 \cup \mathcal{E}_2} = \bar{\mathcal{E}}_1 \cap \bar{\mathcal{E}}_2\,.$$

33. Consider the following matrix $[M]$:

$$[M] = \begin{bmatrix} a & b & b \,\ldots\ldots\, b \\ b & a & b \,\ldots\ldots\, b \\ \vdots & \vdots & \vdots \qquad \vdots \\ b & b & b \,\ldots\ldots\, a \end{bmatrix}.$$

in which all the elements are equal to b except the diagonal ones, which are equal to a. Show that the sum and product of two matrices of this form has the same form.

34. Show that if $[A]$ is a symmetric matrix and $[C]$ an arbitrary matrix, then $[C'AC]$ is symmetric.

35. If the matrix $[A]$ is positive definite, so are the matrices $[A']$ and $[A]^{-1}$.

36. Show that if $[P]$ is a nonsingular square matrix, then $[PP']$ is positive definite.

37. If the matrix $[M]$ is skew-symmetric and the matrix $[I+M]$ [1] is non-singular, the matrix $[I-M]\,[I+M]^{-1}$ is orthogonal.

38. Apply this result to the following matrix $[M]$:

$$[M] = \begin{bmatrix} 0 & -1 & 5 \\ 1 & 0 & 4 \\ -5 & -4 & 0 \end{bmatrix}.$$

39. A square matrix $[M]$ is said to be idempotent if $[M]^2 = [M]$. Show that if $[M]$ is idempotent the same is true of the matrix $[I - M]$, where $[I]$ denotes the unit matrix.

40. Given the well-known relation

$$\sum_1^n x_i^2 = \sum_1^n (x_i - \bar{x})^2 + n\bar{x}^2 ,$$

where

$$\bar{x} = \frac{1}{n}\sum_1^n x_i ,$$

find the matrices of the three quadratic forms figuring in this expression:

$$Q_1 = \sum x_i^2 ,$$
$$Q_2 = \sum (x_i - \bar{x})^2 ,$$
$$Q_3 = n\bar{x}^2 .$$

Find the ranks of these quadratic forms, show that their respective matrices are idempotent and show that Cochran's Theorem is applicable to them.

---

(1) We do not distinguish between $[I+M]$ and $[I]+[M]$.

1. This is proved by a counterexample. Consider two subspaces $\mathcal{E}_1$ and $\mathcal{E}_2$ of E_n and two vectors V_1 and V_2 of $\mathcal{E}_1 \cup \mathcal{E}_2$; $V_1 + V_2$ belongs to $\mathcal{E}_1 \cup \mathcal{E}_2$ only if both V_1 and V_2 are in $\mathcal{E}_1$ or both V_1 and V_2 are in $\mathcal{E}_2$. When $V_1 \in \mathcal{E}_1$ and $V_2 \in \mathcal{E}_2$, $V_1 + V_2$ is not in $\mathcal{E}_1 \cup \mathcal{E}_2$. Thus the set of vectors $V \in \mathcal{E}_1 \cup \mathcal{E}_2$ is not closed under the operation + and so is not a vector space.

On the other hand, it is clear that in the particular case $\mathcal{E}_1 \subset \mathcal{E}_2$ or $\mathcal{E}_2 \subset \mathcal{E}_1$, the union $\mathcal{E}_1 \cup \mathcal{E}_2$ is a vector space.

2. The set of free vectors of ordinary real euclidean space is a vector space is a vector space over the field of real numbers. This is easily seen by verifying the properties I-1 through I-21 for the free vectors of the space E_3.

3.

$$[A] + [B] = \begin{bmatrix} 3 & 6 & 13 & 2 \\ 8 & 3 & 6 & 5 \\ 8 & 17 & 9 & 7 \end{bmatrix}.$$

4.

$$[C] + [D] = \begin{bmatrix} 21 & 6 & 18 & 11 & 13 \\ 3 & 22 & 23 & 14 & 14 \end{bmatrix}$$

5. The matrices $[A]$ and $[B]$ represent linear mappings of E_4 into E_3; $[C]$ and $[D]$ represent linear mappings of E_5 into E_2.

6.

$$[E]\,[F] = \begin{bmatrix} 82 & 42 & 30 & 10 & 32 \\ 13 & 7 & 2 & 1 & 5 \\ 127 & 63 & 63 & 19 & 50 \\ 97 & 49 & 41 & 13 & 38 \end{bmatrix}.$$

This product represents a mapping of E_4 into E_5.

7.

$$[G]\,[H] = \begin{bmatrix} 0 & 27 & 27 & 9 \\ 0 & 15 & 15 & 5 \\ 0 & 9 & 9 & 3 \\ 0 & 21 & 21 & 7 \end{bmatrix}.$$

$$[H]\,[G] = [31].$$

The first matrix represents a mapping of E_4 into itself. The second represents a scalar.

8. First assume that the p vectors U_i are linearly independent. Then

$$f(X) = \sum_1^p \alpha_i(X)\, U_i\,.$$

The mapping f is linear if and only if:

$$f(X + Y) = \sum_1^p \alpha_i(X + Y)\, U_i = f(X) + f(Y) = \sum_1^p \alpha_i(X)\, U_i + \\ + \sum_1^p \alpha_i(Y)\, U_i = \sum_1^p [\alpha_i(X) + \alpha_i(Y)]\, U_i$$

$$f(aX) = \sum_1^p \alpha_i(aX)\, U_i = af(X) = a\sum_1^p \alpha_i(X)\, U_i\ ,$$

and, since the vectors U_i are linearly independent, this implies that

$$\left.\begin{array}{l}\alpha_i(X+Y)=\alpha_i(X)+\alpha_i(Y)\\ \alpha_i(aX) \qquad + a\alpha_i(X)\end{array}\right\} i=1,2,\dots,p\,.$$

In other words, the mapping f is linear only if the functions $\alpha_i(X)$ are linear. When this is so, we have a linear mapping of E_m into a p-dimensional subspace of E_m. The corresponding matrix thus has p rows and m columns. If the vectors U_i are not linearly independent, but constitute a system of rank r, it is easily seen that f may be expressed in the following form:

$$f(X)=\sum_{i=1}^{r}\left[\alpha_i(X)+\sum_{j=r+1}^{p}\alpha_j(X)\,\lambda_i^j\right]U_i$$

where the coefficients λ_j^i do not depend on X; this brings us back to the previous case. But then we have a linear mapping of E_m into an r-dimensional vector subspace, and the corresponding matrix has r rows and m columns.

9. The result is immediate. In fact:

$$\{\,[A]\,[A]'\,\}'=[A]''\,[A]'=[A]\,[A]'$$

and the resulting matrix has m rows and m columns.

10. We have

$$[A]\,[B]=\begin{bmatrix}46 & 64 & 84\\ 23 & 28 & 33\\ 53 & 64 & 75\end{bmatrix}\quad [B]\,[A]=\begin{bmatrix}21 & 19 & 22\\ 66 & 52 & 49\\ 111 & 85 & 76\end{bmatrix}.$$

It is clear that these two matrices are different.

11. Consider the two diagonal matrices A and B. The general element of the matrix $[A][B]$, say p_i^k, is given by the usual formula

$$p_i^k=\sum_j a_i^j\,b_j^k\,.$$

Since the off-diagonal terms a_i^j and b_j^k are zero, we get $p_i^k = 0$ for $i \neq k$, since all products figuring in the sum have at least one zero factor and $p_i^i = a_i^i b_i^i$. This result can be stated as follows: The product of two diagonal matrices is a diagonal matrix whose diagonal elements are the products of the corresponding diagonal elements of the factor-matrices. The result extends immediately to scalar matrices:

The product of two scalar $n \times n$ matrices is a scalar $n \times n$ matrix.

12. Using a previous result, we can prove that the transpose of the sum of two matrices is the sum of their transposes. In fact:

$$[A] + [B] = [a_i^j + b_i^j]$$

$$[[A] + [B]]' = [a_j^i + b_j^i] = [A]' + [B]' .$$

In this case, therefore:

$$[[A] + [A']] = [A]' + [A]'' = [A]' + [A]$$

and since addition of matrices is commutative, we finally obtain

$$[[A] + [A]']' = [A] + [A]' ,$$

and the matrix $[A] + [A]'$ is indeed symmetric.

The general term of the matrix $[A] - [A']$ is:

$$s_i^j - a_i^j = a_j^i$$

and the general term of its transpose is:

$$t_i^j = a_j^i - a_i^j = -(a_i^j - a_j^i) = -s_i^j .$$

Thus the matrix $[A]-[A]'$ is indeed skew-symmetric.

13. We first show that

$$q \leqslant \min r, s\,.$$

If $r < s$ the result is immediate. Indeed, the range D_B of the first mapping $[B]$ is of dimension s. The images of the vectors of D_B under the mapping $[A]$ are in the image D_A of the latter, and so they belong to a subspace of dimension at most r; this proves the relation.

If $r > s$, choose p arbitrary vectors $V_1, V_2, \ldots, V_p$ in D_{AB}, where $s < p \leqslant r$, and consider a linear combination of these vectors:

$$L = \alpha_1 V_1 + \alpha_2 V_2 + \cdots + \alpha_p V_p\,.$$

Since $V_i \in D_{AB}$, we have

$$V_i = [A]\,U_i\,, \quad \text{where} \quad U_i \in D_B\,.$$

Thus

$$L = \sum_1^p \alpha_i V_i = \sum_1^p \alpha_i [A]\,U_i = [A]\left(\sum_1^p \alpha_i U_i\right).$$

But there exist at most s linearly independent vectors in D_B, and so one can find coefficients α_i, not all zero, such that $\sum_1^n \alpha_i U_i$ is zero and so $L = 0$. Thus there are at most s linearly independent vectors V_i.

We now show that

$$q \geqslant r + s - n\,.$$

Decompose E_n into two disjoint subsets, D_B and its complement $\overline{D}_B$ in E_n.

Since D_B is of dimension s, it contains at most s linearly independent vectors. We obtain a basis of E_n by completing these s vectors with $n - s$ linearly independent vectors which must be elements of $\overline{D}_B$.

When $[A]$ is applied to the space E_n, the s basis vectors from D_B are mapped onto the q basis vectors of D_{AB} (by definition). As to the $n - s$ basis vectors from $\overline{D}_B$, these are mapped onto at most $n - s$ basis vectors in D_A. Since the latter is of dimension r, we have:

$$r \leqslant q + (n - s),$$

or

$$q \geqslant r + s - n.$$

14. Consider the determinant:

$$D = D(C_1, C_2, \ldots, C_i, \ldots, C_n)$$

and replace the column C_i by the following column C_i':

$$C_i' = C_i + \sum_{\substack{j=1 \\ j \neq i}}^{n} \alpha_j C_j .$$

Obviously,

$$D(C_1, C_2, \ldots, C_i', \ldots, C_n) = D + \sum_{\substack{j=1 \\ j \neq i}}^{n} \alpha_j D(C_1, C_2, \ldots, C_j, \ldots, C_n) = D$$

since all the determinants in the sum have two identical columns and therefore vanish.

15. The reasoning is analogous to that of the preceding exercise, applied to the row vectors.

16. We first describe a general method, and then indicate how to simplify it in certain particular cases. Consider the determinant:

$$\Delta = \begin{vmatrix} a_1^1 & a_1^2 & \cdots & a_1^n \\ a_2^1 & a_2^2 & \cdots & a_2^n \\ \vdots & & & \\ a_n^1 & a_n^2 & \cdots & a_n^n \end{vmatrix} .$$

The computation algorithm is as follows:

1. Divide the first row of the determinant by a_1^1, to give a determinant Δ':

$$\Delta = a_1^1\,\Delta' = a_1^1 \begin{vmatrix} 1 & a_1^n/a_1^1 & \cdots & a_1^n/a_1^1 \\ a_2^1 & a_2^2 & \cdots & a_2^n \\ \vdots & & & \\ a_n^1 & a_n^2 & & a_n^n \end{vmatrix}$$

2. Subtract the first row, multiplied by a_2', from the second row. The result is a determinant Δ_2' equal to Δ', whose second row is:

$$0 \quad a_2^2 - a_1^2\,a_2^1/a_1^1 \quad \cdots \quad a_2^n - a_1^n\,a_2^1/a_1^1 .$$

3. Subtract the first row of Δ_2', multiplied by a_3', from its third row. The result is a determinant Δ_3' equal to Δ', whose third row is:

$$0 \quad a_3^2 - a_1^2\,a_3^1/a_1^1 \quad \cdots \quad a_3^n - a_1^n\,a_3^1/a_1^1$$

and so on.

After $n - 1$ operations we obtain a determinant Δ_n' equal to Δ':

$$\Delta'_n = \Delta' = \begin{vmatrix} 1 & a_1^2/a_1^1 & \dots & a_1^n/a_1^1 \\ 0 & a_2^2 - a_1^2\, a_2^1/a_1^1 & \dots & a_2^n - a_1^n\, a_2^1/a_1^1 \\ \vdots & & & \\ 0 & a_n^2 - a_1^2\, a_n^1/a_1^1 & \dots & a_n^n - a_1^n\, a_n^1/a_1^1 \end{vmatrix}.$$

Now, expanding Δ'_n in terms of its first column, we get a determinant of order $n - 1$.

The computations can sometimes be simplified when the coefficients of the determinant allow this, as evidenced by the following example.

17. For convenience' sake, let us call the five rows of the determinant $L_1, L_2, \dots, L_5$. Apply the following transformations, which do not change the value of the determinant, in succession:

1) $L_5 \to L_5 - L_1 = L'_5$,

2) $L_2 \to L_2 - L_1 - 2\,L_4 = L'_2$,

3) $L_3 \to L_3 - 2\,L_1 - L_4 = L'_3$,

4) $L_1 \to L_1 - 3\,L_4 = L'_1$.

After these four operations we get:

$$\Delta = \begin{vmatrix} 0 & -8 & -5 & -25 & -11 \\ 0 & -5 & -13 & -16 & -11 \\ 0 & -2 & -13 & -5 & -11 \\ 1 & 3 & 4 & 9 & 5 \\ 0 & 0 & -5 & 5 & -1 \end{vmatrix} = \begin{vmatrix} 8 & 5 & 25 & 11 \\ 5 & 13 & 16 & 11 \\ 2 & 13 & 5 & 11 \\ 0 & -5 & 5 & -1 \end{vmatrix}.$$

Now let $\Lambda_1, \Lambda_2, \dots, \Lambda_4$ be the rows of this new determinant, and apply the following transformations:

1) $\Lambda_1 \to \Lambda_1 - \Lambda_2 = \Lambda'_1$,
2) $\Lambda_2 \to \Lambda_2 - \Lambda_3 = \Lambda'_2$,
3) $\Lambda_3 \to \Lambda_3 + 11\,\Lambda_4 = \Lambda'_4$.

The result is:

$$\Delta - \begin{vmatrix} 3 & -8 & 9 & 0 \\ 3 & 0 & 11 & 0 \\ 2 & -42 & 60 & 0 \\ 0 & -5 & 5 & -1 \end{vmatrix} = - \begin{vmatrix} 3 & -8 & 9 \\ 3 & 0 & 11 \\ 2 & -42 & 60 \end{vmatrix} = -1516 .$$

18. Carry out the following transformations in succession:

1) $L_6 \to L_6 - L_1 = L'_6,$
2) $C_4 \to C_4 - C_6 = C'_4,$
3) $C_2 \to C_2 + 4\,C_6 = C'_2,$

and the result is:

$$\Delta = \begin{vmatrix} 3 & 37 & 1 & -2 & 2 & 7 \\ 2 & 4 & 5 & 3 & 8 & 1 \\ 6 & 17 & 5 & -1 & 9 & 4 \\ 8 & 16 & 0 & 4 & 3 & 3 \\ 5 & 24 & 2 & -4 & 7 & 4 \\ 0 & 0 & 0 & 0 & 0 & 1 \end{vmatrix} = \begin{vmatrix} 3 & 37 & 1 & -2 & 2 \\ 2 & 4 & 5 & 3 & 8 \\ 6 & 17 & 5 & -1 & 9 \\ 8 & 16 & 0 & 4 & 3 \\ 5 & 24 & 2 & -4 & 7 \end{vmatrix} .$$

Apply the following transformations to this new determinant:

1) $L_5 \to L_5 - L_1 - L_2 = L'_5,$
2) $L_4 \to L_4 - L_2 - L_3 = L'_4,$
3) $L_3 \to L_3 - 2\,L_1 = L'_3,$
4) $L_1 \to L_1 - L_2 = L'_1,$
5) $L_2 \to L_2 - 2\,L'_1,$

and we get:

$$\Delta = \begin{vmatrix} 1 & 34 & -4 & -5 & -6 \\ 0 & -64 & 13 & 13 & 20 \\ 0 & -57 & 3 & 3 & 5 \\ 0 & -5 & -10 & 2 & -14 \\ 0 & -17 & -4 & -5 & -3 \end{vmatrix} = \begin{vmatrix} -64 & 13 & 13 & 20 \\ -57 & 3 & 3 & 5 \\ -5 & -10 & 2 & -14 \\ -17 & -4 & -5 & -3 \end{vmatrix} .$$

Finally, apply the following transformations to the new determinant:

$$1)\quad L_1 \to L_1 - 4\,L_2 = L'_1,$$
$$2)\quad C_3 \to C_3 - C_2 = C'_3,$$
$$3)\quad C_1 \to C_1 - 164\,C_2 = C'_1,$$

and the result is

$$\Delta = \begin{vmatrix} 0 & 1 & 0 & 0 \\ -549 & 3 & 0 & 5 \\ 1\,635 & -10 & 12 & -14 \\ 639 & -4 & -1 & -3 \end{vmatrix} = \begin{vmatrix} -549 & 0 & 5 \\ 1\,635 & 12 & -14 \\ 639 & -1 & -5 \end{vmatrix} - 32\,241\,.$$

19. Applying the classical formulas, we get:

$$[A]\,[B] = \begin{bmatrix} 28 & 44 & 28 \\ 33 & 111 & 78 \\ 29 & 63 & 40 \end{bmatrix}$$

and

$$|\,AB\,| = -\,3\,744\,.$$

On the other hand,

$$|\,A\,| = -\,26\,, \qquad |\,B\,| = 144$$

and so

$$|\,A\,|\,|\,B\,| = -\,26 \times 144 = -\,3\,744\,.$$

Thus the equality is proved.

20. The conjugate permutations are:

2 6 3 1 4 9 5 8 7
3 4 2 5 8 9 7 1 6
7 5 1 6 4 8 2 9 3
5 3 4 7 1 8 2 9 6
1 3 5 8 2 6 7 9 4
6 9 3 7 5 2 1 8 4.

21. The numbers of inversions in these permutations are: 10 ; 14 ; 18 ; 14 ; 11 ; 23.

22. The maximum number of inversions is obtained when each element forms an inversion with each of the following terms. This permutation is:

$$n(n-1)\dots 1 .$$

There is only one such permutation, and the number of inversions is

$$i_n = \sum_1^n n - 1 = \frac{n(n-1)}{2} .$$

In the case $n = 9$,

$$i_9 = \frac{9 \times 8}{2} = 36 .$$

23. The determinant of the system is:

$$\begin{vmatrix} 1 & 1 & 1 \\ 1 & 1 & 0 \\ 0 & 1 & 1 \end{vmatrix} = 1$$

and it is not zero. We thus have a Cramer system, which has the following solution:

$$x = -2, \quad y = 4, \quad z = -1 .$$

24. The determinant of the system,

$$\begin{vmatrix} 1 & 1 & 1 \\ 1 & 0 & 2 \\ 0 & 1 & -1 \end{vmatrix}$$

is zero, and so this is not a Cramer system. One immediately verifies that the third equation is the difference of the first two.

The system reduces to two equations:

$$\begin{aligned} x + y + z &= 1, \\ x + 2z &= 2, \end{aligned}$$

which have the following solution (involving z as a parameter):

$$\begin{aligned} x &= 2(1 - z), \\ y &= z - 1. \end{aligned}$$

25. The determinant of the system is:

$$\Delta = \begin{vmatrix} \alpha & 1 & -1 \\ 1 & -\alpha & 1 \\ -1 & 1 & \alpha \end{vmatrix} = -(\alpha^3 + \alpha + 2) = -(\alpha + 1)(\alpha^2 - \alpha + 2)$$

The second factor vanishes only for imaginary values of α, and so this is a Cramer system for any real α other than -1. Under these conditions ($\alpha \neq -1$):

$$x = \frac{\begin{vmatrix} 1 & 1 & -1 \\ 2\alpha & -\alpha & 1 \\ 1 & 1 & \alpha \end{vmatrix}}{\Delta} = \frac{3\alpha^2 + 3\alpha + 2}{(\alpha + 1)(\alpha^2 - \alpha + 2)},$$

$$y = \frac{\begin{vmatrix} \alpha & 1 & -1 \\ 1 & 2\alpha & 1 \\ -1 & 1 & \alpha \end{vmatrix}}{\Delta} = -2\,\frac{\alpha^3 + 4\alpha + 2}{(\alpha + 1)(\alpha^2 - \alpha + 2)},$$

$$z = \frac{\begin{vmatrix} \alpha & 1 & 1 \\ 1 & -\alpha & 2\alpha \\ -1 & 1 & 1 \end{vmatrix}}{\Delta} = \frac{3\alpha}{\alpha^2 - \alpha + 2}.$$

When $\alpha = -1$ the system is degenerate:

$$\begin{aligned} -x + y - z &= 1, \\ x + y + z &= -2, \\ -x + y - z &= 1. \end{aligned}$$

The first and third equations coincide and, regarding z as a parameter, we write the system as:

$$\begin{aligned} -x + y &= 1 + z, \\ x + y &= -2 - z, \end{aligned}$$

so that

$$x = -\frac{3}{2} - z, \quad y = -\frac{1}{2}.$$

26. The system involves 3 unknowns and four equations and is of order at most 3. For the principal determinant, take:

$$\Delta = \begin{vmatrix} \alpha & 1 & 1 \\ 1 & \alpha & 1 \\ 1 & 1 & \alpha \end{vmatrix} = (\alpha - 1)^2 (\alpha + 2),$$

When $\alpha = 1$, the system is of order 1, since all the minors of order 2 vanish. The four equations reduce to a single one, and the system is consistent if

$$A = B = C = 1.$$

In this case the system depends on two parameters, say y and z.

When $\alpha = -2$, the above determinant vanishes, but the matrix of the system contains a nonzero determinant of order 3; for example:

$$\Delta' = \begin{vmatrix} 1 & -2 & 1 \\ 1 & 1 & =2 \\ 1 & 1 & 1 \end{vmatrix} -9 .$$

Thus the system is of order 3 and has a solution if and only if its characteristic determinant vanishes, i.e.:

$$D_c = \begin{vmatrix} -2 & 1 & 1 & A \\ 1 & -2 & 1 & B \\ 1 & 1 & -2 & C \\ 1 & 1 & 1 & 1 \end{vmatrix} = \begin{vmatrix} 0 & 0 & 0 & A+B+C \\ 1 & -2 & 1 & B \\ 1 & 1 & -2 & C \\ 1 & 1 & 1 & 1 \end{vmatrix}$$

$$= 9(A + B + C) .$$

Assume that

$$A + B + C = 0 .$$

The system then becomes:

$$\begin{aligned} -2x + y + z &= A , \\ x - 2y + z &= B , \\ x + y + z &= 1 , \end{aligned}$$

and this Cramer system has the solution:

$$x = \frac{1 - A}{3}, \quad y = \frac{1 - B}{3}, \quad z = \frac{1 + A + B}{3} .$$

In the general case, when α is neither 1 nor -2, the system has a solution when its characteristic determinant vanishes, that is, when:

$$\Delta = \begin{vmatrix} \alpha & 1 & 1 & A \\ 1 & \alpha & 1 & B \\ 1 & 1 & \alpha & C \\ 1 & 1 & 1 & 1 \end{vmatrix} = 0,$$

$$\Delta_c = -(\alpha - 1)^2 [A + B + C - \alpha + 2],$$

i.e.,

$$A + B + C = \alpha + 2.$$

Under these conditions, the system reduces to its first three equations:

$$\begin{aligned} \alpha x + y + z &= A, \\ x + \alpha y + z &= B, \\ x + y + \alpha z &= C, \end{aligned}$$

and we find:

$$x = \frac{\begin{vmatrix} A & 1 & 1 \\ B & \alpha & 1 \\ C & 1 & \alpha \end{vmatrix}}{\Delta} = \frac{-A(1+\alpha) + B + C}{2+\alpha}$$

$$y = \frac{\begin{vmatrix} \alpha & A & 1 \\ 1 & B & 1 \\ 1 & C & \alpha \end{vmatrix}}{\Delta} = \frac{A - B(1+\alpha) + C}{2+\alpha},$$

$$z = \frac{\begin{vmatrix} \alpha & 1 & A \\ 1 & \alpha & B \\ 1 & 1 & C \end{vmatrix}}{\Delta} = \frac{A + B - C(1+\alpha)}{2+\alpha}.$$

27. Direct computation in the first reference system gives:

$$f(X, Y) = X'[M] Y,$$

where

$$[M]\,Y = \begin{bmatrix} -2 \\ 15 \\ 29 \end{bmatrix} \text{ and } X'[M]\,Y = [1\ 2\ 3] \begin{bmatrix} -2 \\ 5 \\ 29 \end{bmatrix} = 115\,.$$

If we go over to the reference system defined by the matrix $[B]$, we get:

$$X_B = [B]^{-1}\,X_A\,,$$
$$Y_B = [B]^{-1}\,Y_A\,,$$

where

$$[B]^{-1} = \begin{bmatrix} 19/8 & -\ 58/8 & 3/8 \\ 1/8 & -\ 6/8 & 1/8 \\ -\ 11/8 & 34/8 & -\ 3/8 \end{bmatrix},$$

and this gives

$$X_B = \begin{bmatrix} -\ 11 \\ -\ 1 \\ 6 \end{bmatrix} \quad Y_B - \begin{bmatrix} 37/8 \\ -\ 1/8 \\ -\ 21/8 \end{bmatrix}.$$

But, on the other hand,

$$[M^B] = [B]'\,[M]\,[B]\,,$$

or

$$[M^B] = \begin{bmatrix} 115 & 294 & 237 \\ 73 & 167 & 142 \\ 206 & 523 & 423 \end{bmatrix}.$$

Hence

$$[M^B]\,Y_B = \begin{bmatrix} -\ 1\,016/8 \\ -\ 448/8 \\ -\ 1\,784/8 \end{bmatrix} = \begin{bmatrix} -\ 127 \\ -\ 56 \\ -\ 223 \end{bmatrix}$$

and

$$X'_B[M^B]\,Y_B = 115\,.$$

28. *The condition is necessary.*

$$X'_A[M_A]\,Y_A = X'_B[M_B]\,Y_B\,.$$

But:

$$X_A = [B]\,X_B\,,\qquad Y_A = [B]\,Y_B\,,\qquad M_A = [B]\,[M_B]\,[B]^{-1}\,.$$

Substitute these values into the right-hand side of the first equality:

$$X'_B[B]'\,[B]\,[M_B]\,[B]^{-1}\,[B]\,Y_B = X'_B[B]'\,[B]\,[M_B]\,Y_B = X'_B[M_B]\,Y_B\,,$$

which implies:

$$[B]'\,[B] = [I]\,.$$

Thus the matrix $[B]$ is orthogonal and the change of basis that it represents is orthogonal.

The condition is sufficient.

Since the change of basis is orthogonal, the representing matrix is also orthogonal and:

$$[B]'\,[B] = [I]\,.$$

On the other hand, using the formulas for change of basis, we get:

$$X'_A[M_A]\,Y_B = X'_B[B]'\,[B]\,[M_B]\,[B]^{-1}\,[B]\,Y_B = X'_B[M_B]\,Y_B\,.$$

29. Replacing the vector X by $X_1 + X_2$ and then by αX, we immediately see that the mapping $f_V(X)$ is linear, of rank 1.

As for the set of all mappings f_V, define the sum of mappings f_{V_1} and f_{V_2} to be the mapping S such that:

$$S(X) = V_1'[M]\,X + V_2'[M]\,X\,.$$

Multiplication by a scalar is defined in a similar way, as the mapping P such that

$$P(X) = \alpha f_V(X) = \alpha V'[M]\,X\,.$$

It is now easy to verify that the set $\mathcal{E}$ of mappings f_V is a vector space, since all the relations I-12 through I-21 are satisfied. For example, let us show that the set is closed under the operation +:

$$\begin{aligned} S(X) &= V_1'[M]\,X + V_2'[M]\,X = (V_1' + V_2')\,[M]\,X \\ &= (V_1 + V_2)'\,[M]\,X\,. \end{aligned}$$

Now the vectors V_1 and V_2 belong to E_n, and so the same holds for their sum and the second term in the above expression defines an element of $\mathcal{E}$.

Considering a linear combination of mappings f_V, we again easily see that the space is of dimension n, since one can find n linearly independent mappings' but not $n + 1$.

30. Write down the determinant of the vectors:

$$D = \begin{vmatrix} 4 & 3 & 1 \\ 0 & 3 & 4 \\ 3 & 5 & 7 \end{vmatrix} = 31\,.$$

Since the determinant is not zero, the vectors X_1, X_2 and X_3 constitute a basis of E_3.

The vector e_1 is:

$$e_1 = \frac{X_1}{\sqrt{X_1'\,X_1}} = \left(\frac{4}{5}, 0, \frac{3}{5}\right).$$

Now set:

$$e_2 = \frac{X_2 - (e_1' X_2)\, e_1}{|X_2 - (e_1' X_2)\, e_1|}.$$

But

$$e_1' X_2 = 27/5$$

and so

$$X_2 - (e_1' X_2)\, e_1 = \begin{bmatrix} -33/25 \\ 3 \\ 44/25 \end{bmatrix}$$

and

$$e_2 = \left(-\frac{33}{\sqrt{8\,650}}, \frac{75}{\sqrt{8\,650}}, \frac{44}{\sqrt{8\,650}}\right).$$

Similarly,

$$e_3 = \frac{X_3 - (e_1' X_3)\, e_1 - (e_2' X_3)\, e_2}{|X_3 - (e_1' X_3)\, e_1 - (e_2' X_3)\, e_2|},$$

and so:

$$e_3 = \left(-\frac{279}{\sqrt{334.211}}, -\frac{341}{\sqrt{334.211}}, \frac{372}{\sqrt{334.211}}\right).$$

It is easily seen that the vectors e_1, e_2 and e_3 are orthogonal and of unit length.

31. We first verify that the four vectors A, B, C and D constitute a basis, by computing the determinant with column vectors A, B, C and D. The value of this determinant is: $\Delta = 209$.

Thus the vectors are indeed a basis of E_4.

Now take:

$$e_1 = \frac{A}{|A|}, \quad \text{where } A| = \sqrt{39},$$

that is:

$$e_1 = (0.160 , 0.480 , 0.801 , 0.320) .$$

Now take:

$$e_2 = \frac{\varepsilon_2}{|\varepsilon_2|},$$

where

$$\varepsilon_2 = B - (e_1' B) e_1 ,$$

The components of this vector are:

$$\begin{aligned} \varepsilon_2^1 &= 5 - 2.561 \cdot 0.160 \\ \varepsilon_2^2 &= 0 - 2.561 \cdot 0.480 \text{ since } C_1' B = 2{,}561 \\ \varepsilon_2^3 &= 1 - 2.561 \cdot 0.801 \\ \varepsilon_2^4 &= 3 - 2.561 \cdot 0.320 \end{aligned}$$

and we get:

$$e_2 = (0.861 , - 0.230 , - 0.197 , 0.409) .$$

Similarly,

$$e_3 = (0.322 , 0.770 , - 0.360 , - 0.417) ,$$

$$e_4 = (- 0.342 , 0.331 , - 0.411 , 0.710) .$$

It is easily seen that the vectors e_1, e_2, e_3 and e_4 constitute an orthonormal system, by verifying that the matrix $[G]$ of these vectors is orthogonal, i.e. $[G'][G] = I$:

$$[G]'\,[G] = \begin{bmatrix} 0.160 & 0.480 & 0.801 & 0.320 \\ 0.861 & -0.230 & -0.197 & 0.409 \\ 0.322 & 0.770 & -0.360 & -0.417 \\ -0.364 & 0.351 & -0.436 & 0.744 \end{bmatrix} \begin{bmatrix} 0.160 & 0.861 & 0.322 & -0.364 \\ 0.480 & -0.230 & 0.770 & 0.351 \\ 0.801 & -0.197 & -0.360 & -0.436 \\ 0.320 & 0.409 & -0.417 & 0.744 \end{bmatrix}$$

which gives

$$[G]'\,[G] = \begin{bmatrix} 1.0000 & 0.0004 & 0.0007 & -0.0009 \\ 0.0004 & 1.0003 & 0.0005 & 0.0040 \\ -0.0007 & 0.0005 & 1.0001 & -0.003 \\ -0.0009 & 0.0044 & -0.002 & 1.0008 \end{bmatrix}.$$

Up to round-off errors, this is the unit matrix.

32. Recall that if $\mathcal{E}$ is a subspace of E_n the orthogonal complement of $\mathcal{E}$ with respect to E_n is the set $\bar{\mathcal{E}}$ of the vectors in E_n orthogonal to $\mathcal{E}$.

We prove first that:

$$\overline{\mathcal{E}_1 \cup \mathcal{E}_2} \subset \bar{\mathcal{E}}_1 \cap \bar{\mathcal{E}}_2,$$

and then that:

$$\bar{\mathcal{E}}_1 \cap \bar{\mathcal{E}}_2 \subset \overline{\mathcal{E}_1 \cup \mathcal{E}_2},$$

which will show that these two sets are identical.

Every vector of $\overline{\mathcal{E}_1 \cup \mathcal{E}_2}$ is orthogonal to the vectors of $\mathcal{E}_1$ and to the vectors of $\mathcal{E}_2$; it therefore belongs to the intersection of $\bar{\mathcal{E}}_1$ and $\bar{\mathcal{E}}_2$ and this proves the first relation.

Every vector of $\bar{\mathcal{E}}_1 \cup \bar{\mathcal{E}}_2$ is orthogonal to the vectors of $\mathcal{E}_1$ and to the vectors of $\mathcal{E}_2$; it therefore belongs to the orthogonal complement of the union of $\mathcal{E}_1$ and $\mathcal{E}_2$, which proves the second relation.

33. Consider two matrices of this type:

$$M_1 = \begin{bmatrix} a & b & \dots & b \\ b & a & \dots & b \\ \vdots & \vdots & & \vdots \\ b & b & \dots & a \end{bmatrix}, \quad M_2 = \begin{bmatrix} c & d & \dots & d \\ d & c & \dots & d \\ \vdots & \vdots & & \vdots \\ d & d & \dots & c \end{bmatrix}.$$

The general term of the matrix $[M_1] + [M_2]$ is:

along the principal diagonal: $a + c$
off the principal diagonal: $b + d$.

The general term of the matrix $[M_1][M_2]$ is:

along the principal diagonal: $(n-1)\,bd + ac$
off the principal diagonal: $(n-2)\,bd + ad + bc$.

Both these matrices are of the type in question.

34. Transposing $[C'AC]$ we find:

$$[C'AC]' = [AC]'\,[C']' = [C]'\,[A]'\,[C] = [C]'\,[A]\,[C].$$

35. First consider the matrix $[A]'$. Considering its terms, one easily sees that:

$$[A - \lambda I]' = [A' - \lambda I].$$

Hence, proceeding to determinants:

$$|A - \lambda I| = |A' - \lambda I|,$$

so that the matrix $[A]'$ has the same eigenvalues as $[A]$. Now if the matrix $[A]$ is positive definite all its eigenvalues are positive, which is a necessary and sufficient condition for a matrix to be positive definite. Hence the matrix $[A]'$, all of whose eigenvalues are positive, is also positive definite.

We now proceed to the matrix $[A]^{-1}$. Given an eigenvalue root λ_i of the matrix $[A]$ and the associated eigenvector Λ_i, we have:

$$[A]\,\Lambda_i = \lambda_i\,\Lambda_i\,.$$

Now apply the matrix $[A]^{-1}$ to both sides, to obtain:

$$\Lambda_i = [A]^{-1}\,\lambda_i\,\Lambda_i = \lambda_i[A]^{-1}\,\Lambda_i\,,$$

that is:

$$[A]^{-1}\,\Lambda_i = \frac{1}{\lambda_i}\,\Lambda_i\,.$$

We have thus obtained two results. Any eigenvector Λ_i of $[A]$ is also an eigenvector of $[A]^{-1}$ and the associated eigenvalue is $1/\lambda_i$, the reciprocal of the corresponding eigenvalue of $[A]$. Consequently, the eigenvalues of the matrix $[A]^{-1}$ have the same sign as those of the matrix $[A]$. In this case they are all positive and so the matrix $[A]^{-1}$ is positive definite.

36. We first show that the matrix $[P'\,P]$ is symmetric, for:

$$[P'\,P]' = [P]'\,[P']' = [P']\,[P] = [P'\,P]\,.$$

On the other hand, the quadratic form Q corresponding to the matrix $[P'\,P]$ is written as:

$$Q = X'[P'\,P]\,X = X'[P]'\,[P]\,X = |\,PX\,|^2\,.$$

Thus Q is always nonnegative; but since the matrix $[P]$ is nonsingular the same holds for $[P'\,P]$ and so

$$Q > 0\,.$$

Thus the matrix $[P'\,P]$ is positive definite.

37. We first see that, since the matrix $[I] + [M]$ is nonsingular, the matrix $[R] = [I - M]\,[I + M]^{-1}$ is positive definite. On the other hand, it is immediately clear that:

$$[I - M]' = [I + M]$$

since $[M]$ is skew-symmetric.

To simplify the notation, we set:

$$[I + M] = [A],$$
$$[I - M] = [A]'.$$

We must now show that the matrix $[R] = [A]'[A]^{-1}$ is orthogonal, by verifying that it satisfies the fundamental relation:

$$[R]'[R] = [I].$$

Now:

$$[R]' = [A]^{-1\prime}[A]$$

and

$$\begin{aligned}[R]'[R] &= [A]^{-1\prime}[A][A]'[A]^{-1}\\ &= [A]^{-1\prime}[A'A]'[A]^{-1}\\ &= [A'AA^{-1}]'[A]^{-1}\\ &= [A']'[A]^{-1} = [A][A]^{-1} = [I].\end{aligned}$$

38. Obviously,

$$[I - M] = \begin{bmatrix} 1 & 1 & -5 \\ -1 & 1 & -4 \\ 5 & 4 & 1 \end{bmatrix}$$

$$[I + M] = \begin{bmatrix} 1 & -1 & 5 \\ 1 & 1 & 4 \\ -5 & -4 & 1 \end{bmatrix}.$$

Hence

$$[I + M]^{-1} = \begin{bmatrix} 0.395 & -0.442 & -0.209 \\ -0.488 & 0.605 & 0.023 \\ 0.023 & 0.209 & 0.047 \end{bmatrix}.$$

Finally,

$$[I - M][I + M]^{-1} = \begin{bmatrix} -0.209 & -0.977 & 0.046 \\ -0.884 & 0.209 & 0.419 \\ -0.419 & 0.046 & -0.907 \end{bmatrix}.$$

39. We must show that

$$[I - M]^2 = [I - M].$$

Now

$$\begin{aligned} [I - M]^2 &= [I]^2 - [I][M] - [M][I] + [M]^2 \\ &= [I] - 2[M] + [M]^2 . \end{aligned}$$

But since $[M]$ is idempotent:

$$[M]^2 = [M],$$

and finally:

$$[I - M]^2 = [I - M].$$

40. We shall use the following formula for the terms a_i^k of the matrices of the quadratic forms Q:

$$a_i^k = \frac{1}{2}\frac{\partial^2 Q}{\partial x_i \partial x_k}.$$

Application of this formula to Q_1 gives:

$$\frac{1}{2}\frac{\partial Q_1}{\partial x_i} = x_i, \quad \frac{1}{2}\frac{\partial^2 Q_1}{\partial x_i \partial x_n} = \begin{matrix} 1 \text{ if } k = i. \\ 0 \text{ if } k \neq i. \end{matrix}$$

The matrix of Q_1 is thus precisely the unit matrix $[I]$, which is also obvious from the equality

$$Q_1 = \sum_i x_i^2 = X'X = X'[I]X .$$

We then find:

$$\frac{1}{2}\frac{\partial Q_2}{\partial x_i} = \left(1 - \frac{1}{n}\right)(x_i - \bar{x}) - \frac{1}{n}\sum_{j \neq i}(x_j - \bar{x})$$

$$\frac{1}{2}\frac{\partial^2 Q}{\partial x_i\,\partial x_k} = -\left(1 - \frac{1}{n}\right)\frac{1}{n} - \frac{1}{n}\left(1 - \frac{1}{n}\right) + \frac{1}{n^2}(n-2) = -\frac{1}{n} \qquad k \neq i ,$$

$$\left(1 - \frac{1}{n}\right)^2 + \frac{1}{n^2}(n-1) = 1 - \frac{1}{n} \qquad k = i .$$

Thus:

$$[M_2] = [I] - [M] ,$$

where $[M]$ denotes the matrix all of whose elements are equal to $1/n$.

Finally,

$$\frac{1}{2}\frac{\partial Q_3}{\partial x_i} = \bar{x} , \quad \frac{1}{2}\frac{\partial^2 Q_3}{\partial x_i\,\partial x_k} = \frac{1}{n} .$$

Thus: $[M_3] = [M]$.

It is obvious that the rank of $[I]$ is n, while that of $[M_3]$ is 1.

We now apply the following transformations to the determinant of $[M_2]$:

1. Subtract the n-th column from each of the other columns. The result is a determinant all of whose elements are zero except the diagonal elements, which are equal to 1 (the last element of the diagonal is still $1 - 1/n$), and those of the last row are equal to -1 (and those of the last column are $-1/n$).

2. In the last determinant, subtract the sum of the first $n - 1$ rows from the last row; we get the determinant:

$$D = \begin{vmatrix} 1 & 0 & 0 & \cdots & & -\frac{1}{n} \\ 0 & 1 & 0 & \cdots & & -\frac{1}{n} \\ \vdots & & & & & \vdots \\ 0 & 0 & 0 & & 1 & -\frac{1}{n} \\ 0 & 0 & 0 & & 0 & 0 \end{vmatrix}$$

The determinant is therefore zero, but the minor of the element a_n^n is equal to 1. Thus the rank of the matrix $[M_2]$ is $n - 1$.

The matrix $[I]$ is clearly idempotent, and it is easily seen that this is also true for the matrix $[M_3]$. Taking an arbitrary element p_i^j of the last-named matrix, we see that:

$$p_i^j = \sum_1^n \frac{1}{n} \times \frac{1}{n} = \frac{1}{n}.$$

Thus:

$$[M_3]^2 = [M_3].$$

It follows that the matrix $[M_2]$, which is equal to $[I - M_3]$, is also idempotent (cf. Exercise 39).